AF553428

FUNDAMENTALS OF OPTICAL ENGINEERING

FUNDAMENTALS OF OPTICAL ENGINEERING

By

S. Singh

D P H

DISCOVERY PUBLISHING HOUSE PVT. LTD.
NEW DELHI-110 002

Published by:
Tilak Wasan

DISCOVERY PUBLISHING HOUSE PVT. LTD.
4383/4B, Ansari Road, Darya Ganj
New Delhi-110 002 (India)
Phone : +91-11-23279245, 23253475, 43596065
E-mail : discoverypublishinghouse@gmail.com
sales@discoverypublishinggroup.com
web : www.discoverypublishinggroup.com

***Edition:* 2020**

ISBN: 978-81-8356-436-6

Fundamentals of Optical Engineering

Printed at:
Infinity Imaging Systems
Delhi

Contents

Preface

The discipline of optical engineering is not merely interesting, but it is often quite fascinating. This book is designed to assist the students, who are interested or involved in the field of optical engineering, to obtain a better understanding of the basic principles and to prepare the reader for more complex topics that would be encountered and dealt with, using more advanced and specialised material. One goal of this book is to demonstrate that there are many aspects of the science of optics but the field of optical engineering is not that difficult to understand.

Optical engineering is an engineering discipline, but, it has no particular academic departments for its exclusive teaching. Basically, optical engineering is the field of study that focuses on applications of optics. Optical engineering and metrology use optical methods to measure micro-vibrations, with instruments, like laser speckle, interferometer or to measure the properties of various masses, with instruments, measuring refraction. Optical engineering uses research and development of laser technology, modern photonic detection and optical metrology for engineering applications.

Not many books, on the particular subject; Optical Engineering, are available in the market. Seeing this dearth, we have made this effort, in order to fill the bill. This book is presented with the belief that it would serve to meet the needs and quench the thirst of the readers. Enlightening comments from the concerned reader are solicited.

—Editor

Introduction

Optical engineering is the field of study that focuses on applications of optics. Optical engineers design components of optical instruments such as lenses, microscopes, telescopes, and other equipment that utilise the properties of light. Other devices include optical sensors and measurement systems, lasers, fibre optic communication systems, optical disc systems (e.g. CD, DVD), etc. Since optical engineers want to design and build devices that make light do something useful, they must understand and apply the science of optics in substantial detail, in order to know what is physically possible to achieve (physics and chemistry). However, they also must know what is practical in terms of available technology, materials, costs, design methods, etc. As with other fields of engineering, computers are important to many (perhaps most) optical engineers. They are used with instruments, for simulation, in design, and for many other applications. Engineers often use general computer tools such as spreadsheets and programming languages, and they make frequent use of specialised optical software designed specifically for their field.

Optical engineering metrology uses optical methods to measure micro-vibrations with instruments like the laser speckle interferometer or to measure the properties of the various masses with instruments measuring refraction.

Optics

Optics is the science that describes the behaviour and properties of light and the interaction of light with matter. Optics explains optical phenomena.

The field of optics usually describes the behaviour of visible, infrared, and ultraviolet light; however, because light is an electromagnetic wave, similar phenomena occur in x-rays, microwaves, radio waves, and other forms of electromagnetic radiation and analogous phenomena occur with charged particle beams. Optics can largely be regarded as a sub-field of electromagnetism. Some optical phenomena depend on the quantum nature of light relating some areas of optics to quantum mechanics. In practice, the vast majority of optical phenomena can be accounted for using the electromagnetic description of light, as described by Maxwell's Equations.

The field of optics has its own identity, societies, and conferences. The pure science aspects of the field are often called optical science or optical physics. Applied optical sciences are often called *optical engineering*. Applications of optical engineering related specifically to illumination systems are called *illumination engineering*. Each of these disciplines tends to be quite different in its applications, technical skills, focus, and professional affiliations. More recent innovations in optical engineering are often categorised as photonics or optoelectronics. The boundaries between these fields and "optics" are often unclear, and the terms are used differently in different parts of the world and in different areas of industry.

Because of the wide application of the science of "light" to real-world applications, the areas of optical science and optical engineering tend to be very cross-disciplinary. Optical science is a part of many related disciplines including electrical engineering, physics, psychology, medicine (particularly ophthalmology and optometry), and others. Additionally, the most complete description of optical behaviour, as known to physics, is unnecessarily complicated for most problems, so particular simplified models are used. These limited models adequately describe subsets of optical phenomena while ignoring behaviour irrelevant and/or undetectable to the system of interest.

Classical Optics

Before quantum optics became important, optics consisted mainly of the application of classical electromagnetism and its high frequency approximations to light. Classical optics divides into two main branches: geometric optics and physical optics.

Geometric optics, or *ray optics*, describes light propagation in terms of "rays". Rays are bent at the interface between two dissimilar media, and may be curved in a medium in which the refractive index is a function of position. The "ray" in geometric optics is an abstract object, or "instrument," which is perpendicular to the wavefronts of the actual optical waves. Geometric optics provides rules for propagating these rays through an optical system, which indicates how the actual wavefront will propagate. This is a significant simplification of optics, and fails to account for many important optical effects such as diffraction and polarisation. It is a good approximation, however, when the wavelength is very small compared with the size of structures with which the light interacts. Geometric optics can be used to describe the geometrical aspects of imaging, including optical aberrations.

Geometric optics is often simplified even further by making the paraxial approximation, or "small angle approximation." The mathematical behaviour then becomes linear, allowing optical components and systems to be described by simple matrices. This leads to the techniques of Gaussian optics and paraxial raytracing, which are used to find first-order properties of optical systems, such as approximate image and object positions and magnifications. Gaussian beam propagation is an expansion of paraxial optics that provides a more accurate model of coherent radiation like laser beams. While still using the paraxial approximation, this technique partially accounts for diffraction, allowing accurate calculations of the rate at which a laser beam expands with distance, and the minimum size to which the beam can be focused. Gaussian beam propagation thus bridges the gap between geometric and physical optics.

Physical optics or wave optics builds on Huygens's principle and models the propagation of complex wavefronts through optical systems, including both the amplitude and the phase of the wave. This technique, which is usually applied numerically on a computer, can account for diffraction, interference, and polarisation effects, as well as other complex effects. Approximations are still generally used, however, so this is not a full electromagnetic wave theory model of the propagation of light. Such a full model is much more computationally demanding, but can be used to solve small-scale problems that require this more accurate treatment.

History of Optics

Optics began with the development of lenses by the ancient Egyptians and Mesopotamians, followed by theories on light and vision developed by ancient Greek and Indian philosophers, and the development of geometrical optics in the Greco-Roman world. Optics was significantly reformed with the development of physical optics and physiological optics in the medieval Islamic world, and then significantly advanced in early modern Europe, where diffractive optics began. These earlier studies on optics are now known as "classical optics". The term "modern optics" refers to areas of optical research that largely developed in the 20th century, such as quantum optics.

Early History of Optics

The earliest known lenses were made from polished crystal, often quartz, and have been dated as early as 700 BC for

Assyrian lenses such as the Layard / Nimrud lens. There are many similar lenses from ancient Egypt, Greece and Babylon. The ancient Romans and Greeks filled glass spheres with water to make lenses. However, glass lenses were not thought of until the Middle Ages.

In ancient India, the philosophical schools of Samkhya and Vaisheshika, from around the 6th-5th century BC, developed theories on light. According to the Samkhya School, light is one of the five fundamental "subtle" elements (*tanmatra*) out of which emerge the gross elements.

In contrast, the Vaisheshika School gives an atomic theory of the physical world on the non-atomic ground of ether, space and time. The basic atoms are those of earth (*prithvi*), water (*apas*), fire (*tejas*), and air (*vayu*), that should not be confused with the ordinary meaning of these terms. These atoms are taken to form binary molecules that combine further to form larger molecules. Motion is defined in terms of the movement of the physical atoms. Light rays are taken to be a stream of high velocity of *tejas* (fire) atoms. The particles of light can exhibit different characteristics depending on the speed and the arrangements of the *tejas* atoms. Around the first century BC, the *Vishnu Purana* refers to sunlight as "the seven rays of the sun".

In the fifth century BC, Empedocles postulated that everything was composed of four elements; fire, air, earth and water. He believed that Aphrodite made the human eye out of the four elements and that she lit the fire in the eye which shone out from the eye making sight possible. If this were true, then one could see during the night just as well as during the day, so Empedocles postulated an interaction between rays from the eyes and rays from a source such as the sun.

In 55 BC, Lucretius, a Roman who carried on the ideas of earlier Greek atomists, wrote:

"The light and heat of the sun; these are composed of minute atoms which, when they are shoved off, lose no time in

shooting right across the interspace of air in the direction imparted by the shove."

– *On the Nature of the Universe*

Despite being similar to later particle theories, Lucretius's views were not generally accepted and light was still theorised as emanating from the eye.

Later in 499, Aryabhata, who proposed a heliocentric solar system of gravitation in his *Aryabhatiya*, wrote that the planets and the Moon do not have their own light but reflect the light of the Sun.

The Indian Buddhists, such as Dignaga in the 5th century and Dharmakirti in the 7th century, developed a type of atomism that is a philosophy about reality being composed of atomic entities that are momentary flashes of light or energy. They viewed light as being an atomic entity equivalent to energy, similar to the modern concept of photons, though they also viewed all matter as being composed of these light/ energy particles.

The Beginnings of Geometrical Optics: The early writers discussed here treated vision more as a geometrical than as a physical, physiological, or psychological problem. The first known author of a treatise on geometrical optics was the geometer Euclid (c. 325 BC-265 BC). Euclid began his study of optics as he began his study of geometry, with a set of self-evident axioms:

- Lines (or visual rays) can be drawn in a straight line to the object.
- Those lines falling upon an object form a cone.
- Those things upon which the lines fall are seen.
- Those things seen under a larger angle appear larger.
- Those things seen by a higher ray, appear higher.
- Right and left rays appear right and left.
- Things seen within several angles appear clearer.

Euclid did not define the physical nature of these visual rays but, using the principles of geometry, he discussed the effects of perspective and the rounding of things seen at a distance.

Where Euclid had limited his analysis to simple direct vision, Hero of Alexandria (c. AD 10-70) extended the principles of geometrical optics to consider problems of reflection (catoptrics). Unlike Euclid, Hero occasionally commented on the physical nature of visual rays, indicating that they proceeded at great speed from the eye to the object seen and were reflected from smooth surfaces but could become trapped in the porosities of unpolished surfaces. This has come to be known as *emission theory*.

Hero demonstrated the equality of the angle of incidence and reflection on the grounds that this is the shortest path from the object to the observer. On this basis, he was able to define the fixed relation between an object and its image in a plane mirror. Specifically, the image appears to be as far behind the mirror as the object really is in front of the mirror.

Like Hero, Ptolemy (c. 90-168) considered the visual rays as proceeding from the eye to the object seen, but unlike Hero, considered that the visual rays were not discrete lines, but formed a continuous cone. Ptolemy extended the study of vision beyond direct and reflected vision; he also studied vision by refracted rays (dioptrics), when we see objects through the interface between two media of different density. He conducted experiments to measure the path of vision when we look from air to water, from air to glass, and from water to glass and tabulated the relationship between the incident and refracted rays.

His tabulated results have been studied for the air water interface, and in general the values he obtained reflect the theoretical refraction given by modern theory, but the outliers are distorted to represent Ptolemy's a priori model of the nature of refraction.

Optical Revolution in the Islamic World

Al-Kindi (c. 801-873) was one of the earliest important optical writers in the Islamic world. In a work known in the west as *De radiis stellarum*, al-Kindi developed a theory "that everything in the world... emits rays in every direction, which fill the whole world." This theory of the active power of rays had an influence on later Western scholars such as Robert Grosseteste and Roger Bacon.

Ibn Sahl (c. 940-1000) was a Persian mathematician associated with the court of Baghdad. About 984, he wrote a treatise *On Burning Mirrors and Lenses* in which he set out his understanding of how curved mirrors and lenses bend and focus light. In his work, he discovered a law of refraction mathematically equivalent to Snell's law. He used his law of refraction to compute the shapes of lenses and mirrors that focus light at a single point on the axis.

The Beginnings of Physical Optics: Ibn al-Haytham (known in as *Alhacen* or *Alhazen* in Western Europe) (965-1040), often regarded as the "father of modern optics", formulated "the first comprehensive and systematic alternative to Greek optical theories." He initiated a revolution in optics and visual perception, and laid the foundations for modern physical optics. Ibn al-Haytham's key achievement was twofold: first, to insist that vision only occurred because of rays entering the eye and that rays postulated to proceed from the eye had nothing to do with it; the second was to define the physical nature of the rays discussed by earlier geometrical optical writers, considering them as the forms of light and colour. He developed a camera obscura to demonstrate that light and colour from different candles passed through a single aperture in straight lines, without intermingling at the aperture. He then analysed these physical rays according to the principles of geometrical optics. Ibn al-Haytham also employed the experimental scientific method as a form of demonstration in optics. He wrote many books on optics, most significantly the *Book of Optics* (*Kitab al-Manazir* in Arabic), translated into Latin as the

De aspectibus or *Perspectiva,* which disseminated his ideas to Western Europe and had great influence on the later developments of optics.

Another aspect associated with Ibn al-Haytham's optical research is related to systemic and methodological reliance on experimentation (*i'tibar*) and controlled testing in his scientific inquiries. Moreover, his experimental directives rested on combining classical physics (*'ilm tabi'i*) with mathematics (*ta'alim*; geometry in particular) in terms of devising the rudiments of what may be designated as a hypothetico-deductive procedure in scientific research. This mathematical-physical approach to experimental science supported most of his propositions in *Kitab al-Manazir* (*The Optics*; *De aspectibus* or *Perspectivae*) and grounded his theories of vision, light and colour, as well as his research in catoptrics and dioptrics. His legacy was further advanced through the 'reforming' of his *Optics* by Kamal al-Din al-Farisi (d. ca. 1320) in the latter's *Kitab Tanqih al-Manazir* [*The Revision of* (Ibn al-Haytham's) *Optics*]. The *Book of Optics* established experimentation as the norm of proof in optics, and gave optics a physico-mathematical conception at a much earlier date than the other mathematical disciplines of astronomy and mechanics. The book was influential in both the Islamic world and in Western Europe.

Avicenna (980-1037) agreed with Alhazen that the speed of light is finite, as he "observed that if the perception of light is due to the emission of some sort of particles by a luminous source, the speed of light must be finite." Abu Rayhan al-Biruni (973-1048) also agreed that light has a finite speed, and he was the first to discover that the speed of light is much faster than the speed of sound.

Abu Abd Allah Muhammad ibn Maudh, who lived in al-Andalus during the second half of the 11th century, wrote a work on optics later translated into Latin as *Liber de crepisculis,* which was mistakenly attributed to Alhazen. This was a "short work containing an estimation of the angle of depression of the sun at the beginning of the morning twilight and at the end of the evening twilight, and an attempt to calculate on the basis of this and other data the height of the atmospheric moisture responsible for the refraction of the sun's rays." Through his experiments, he obtained the value of 18°, which comes close to the modern value.

In the late-13th and early-14th centuries, Qutb al-Din 7al-Shirazi (1236-1311) and his student Kamal al-Din al-Farisi (1260-1320) continued the work of Ibn al-Haytham, and they were the first to give the correct explanations for the rainbow phenomenon. Al-Farisi published his findings in his *Kitab Tanqih al-Manazir* [*The Revision of* (Ibn al-Haytham's) *Optics*].

In 1574, Taqi al-Din (1526-1585) wrote the last major Arabic work on optics, entitled *Kitab Nur Hadaqat al-Ibsar wa-Nur Haqiqat al-Anzar* (*Book of the Light of the Pupil of Vision and the Light of the Truth of the Sights*), which contains experimental investigations in three volumes on vision, the light's reflection, and the light's refraction. The book deals with the structure of light, its diffusion and global refraction, and the relation between light and colour.

In the first volume, he discusses "the nature of light, the source of light, the nature of the propagation of light, the formation of sight, and the effect of light on the eye and sight". In the second volume, he provides "experimental proof of the specular reflection of accidental as well as essential light, a complete formulation of the laws of reflection, and a description of the construction and use of a copper instrument for measuring reflections from plane, spherical, cylindrical, and conical mirrors, whether convex or concave." The third volume "analyses the important question of the variations light undergoes while travelling in media having different densities, i.e. the nature of refracted light, the formation of refraction, the nature of images formed by refracted light." He also invented an early rudimentary telescope.

The Beginnings of Physiological Optics: Ibn al-Haytham discussed the topics of medicine and ophthalmology in the anatomical and physiological portions of the *Book of Optics* and in his commentaries on Galenic works. He accurately described the process of sight, the structure of the eye, image formation in the eye and the visual system. He also discovered the underlying principles of Hering's law of equal innervation, vertical horopters and binocular disparity, and improved on the theories of binocular vision, motion perception and horopters previously discussed by Aristotle, Euclid and Ptolemy.

He discussed ocular anatomy, and was the first author to deal with the "descriptive anatomy" and "functional anatomy" of the eye independently. Much of his decriptive anatomy was faithful to Galen's gross anatomy, but with significant differences in his approach. For example, the whole area of the eye behind the iris constitutes what Ibn al-Haytham uniquely called the *uveal sphere,* and his description of the eye was devoid of any teleological or humoural theories associated with Galenic anatomy. He also described the eye as being made up of two intersecting globes, which was essential to his functional anatomy of the eye.

After describing the construction of the eye, Ibn al-Haytham makes his most original anatomical contribution in describing the functional anatomy of the eye as an optical system, or optical instrument. His multiple light-source experiment via a reduction slit with the camera obscura, also known as the lamp experiment, provided sufficient empirical grounds for him to develop his theory of corresponding point projection of light from the surface of an object to form an image on a screen. It was his comparison between the eye and the beam-chamber, or *camera obscura,* which brought about his synthesis of anatomy and optics, giving rise to a new field of optics now known as "physiological optics". As he conceptualised the essential principles of pinhole projection from his experiments with the pinhole camera, he considered image inversion to also occur in the eye, and viewed the pupil as being similar to an aperture. Regarding the process of image formation, however, he incorrectly agreed with Avicenna that the lens was the receptive organ of sight, but correctly hinted at the retina also being involved in the process.

Optics in Medieval Europe

The English bishop, Robert Grosseteste (c. 1175-1253), wrote on a wide range of scientific topics at the time of the origin of the medieval university and the recovery of the works of Aristotle. Grosseteste reflected a period of transition between the Platonism of early medieval learning and the new Aristotelianism, hence he tended to apply mathematics and the Platonic metaphor of light in many of his writings. He has been credited with discussing light from four different perspectives: an epistemology of light, a metaphysics or cosmogony of light, an etiology or physics of light, and a theology of light.

Setting aside the issues of epistemology and theology, Grosseteste's cosmogony of light describes the origin of the universe in what may loosely be described as a medieval "big bang" theory. Both his biblical commentary, the *Hexaemeron* (1230 × 35), and his scientific *On Light* (1235 × 40), took their inspiration from Genesis 1:3, "God said, let there be light", and described the subsequent process of creation as a natural physical process arising from the generative power of an expanding (and contracting) sphere of light.

His more general consideration of light as a primary agent of physical causation appears in his *On Lines, Angles, and Figures* where he asserts that "a natural agent propagates its power from itself to the recipient" and in *On the Nature of Places* where he notes that "every natural action is varied in strength and weakness through variation of lines, angles and figures."

The English Franciscan, Roger Bacon (c. 1214-1294) was strongly influenced by Grosseteste's writings on the importance of light. In his optical writings (the *Perspectiva*, the *De multiplicatione specierum*, and the *De speculis comburentibus*) he cited a wide range of recently translated optical and philosophical works, including those of Alhacen, Aristotle, Avicenna, Averroes, Euclid, al-Kindi, Ptolemy, Tideus, and Constantine the African. Although he was not a slavish imitator, he drew his mathematical analysis of light and vision from the writings of the Arabic writer, Alhacen. But he added to this the neo-Platonic concept, perhaps drawn from Grosseteste, that every object radiates a power (*species*) by which it acts upon nearby objects suited to receive those species. Note that Bacon's optical use of the term "species" differs significantly from the genus / species categories found in Aristotelian philosophy.

Another English Franciscan, John Pecham (died 1292) built on the work of Bacon, Grosseteste, and a diverse range of earlier writers to produce what became the most widely used textbook on optics of the Middle Ages, the *Perspectiva Communis*. His book centred on the question of vision, on how we see, rather than on the nature of light and colour. Pecham followed the model set forth by Alhacen, but interpreted Alhacen's ideas in the manner of Roger Bacon.

Like his predecessors, Witelo (c. 1230-1280) drew on the extensive body of optical works recently translated from Greek and Arabic to produce a massive presentation of the subject entitled the *Perspectiva*. His theory of vision follows Alhacen and he does not consider Bacon's concept of species, although passages in his work demonstrate that he was influenced by Bacon's ideas. Judging from the number of surviving manuscripts, his work was not as influential as those of Pecham and Bacon, yet his importance, and that of Pecham, grew with the invention of printing.

Renaissance and Early Modern Optics

Johannes Kepler (1571-1630) picked up the investigation of the laws of optics from his lunar essay of 1600. Both lunar and solar eclipses presented unexplained phenomena, such as unexpected shadow sizes, the red colour of a total lunar eclipse, and the reportedly unusual light surrounding a total solar eclipse. Related issues of atmospheric refraction applied to all astronomical observations.

Through most of 1603, Kepler paused his other work to focus on optical theory; the resulting manuscript, presented to the Emperor on January 1, 1604, was published as *Astronomiae Pars Optica* (*The Optical Part of Astronomy*). In it, Kepler described the inverse-square law governing the intensity of light, reflection by flat and curved mirrors, and principles of pinhole cameras, as well as the astronomical implications of optics such as parallax and the apparent sizes of heavenly bodies. *Astronomiae Pars Optica* is generally recognised as the foundation of modern optics (though the law of refraction is conspicuously absent).

Willebrord Snellius (1580-1626) was most famous for the law of refraction now known as *Snell's law*. He mathematically formulated the law of refraction that is named after him in 1621.

Rene Descartes (1596-1650) made contributions to the field of optics. He showed by using geometric construction and the law of refraction (also known as Descartes' law) that the angular radius of a rainbow is 42 degrees (i.e. the angle subtended at the eye by the edge of the rainbow and the ray passing from the sun through the rainbow's centre is 42°). He also independently discovered the law of reflection, and his essay on optics was the first published mention of this law.

Christiaan Huygens (1629-1695) wrote several works in the area of optics. These included the *Opera reliqua* (also known as *Christiani Hugenii Zuilichemii, dum viveret Zelhemii toparchae, opuscula posthuma*) and the *Traitbe de la lumiaere*.

Isaac Newton (1643-1727) investigated the refraction of light, demonstrating that a prism could decompose white light into a spectrum of colours, and that a lens and a second prism could recompose the multicoloured spectrum into white light. He also showed that the coloured light does not change its properties by separating out a coloured beam and shining it on various objects. Newton noted that regardless of whether it was reflected or scattered or transmitted, it stayed the same colour. Thus, he observed that colour is the result of objects interacting with already-coloured light rather than objects generating the colour themselves.

This is known as *Newton's theory of colour*. From this work he concluded that any refracting telescope would suffer from the dispersion of light into colours, and invented a reflecting telescope (today known as a *Newtonian telescope*) to bypass that problem. By grinding his own mirrors, using Newton's rings to judge the quality of the optics for his telescopes, he was able to produce a superior instrument to the refracting telescope, due primarily to the wider diameter of the mirror. In 1671, the Royal Society asked for a demonstration of his reflecting telescope. Their interest encouraged him to publish his notes *On Colour*, which he later expanded into his *Opticks*. Newton argued that light is composed of particles or *corpuscles* and were refracted by accelerating towards the denser medium, but he had to associate them with waves to explain the diffraction of light (*Opticks* Bk. II, Props. XII-L). Later physicists instead favoured a purely wavelike explanation of light to account for diffraction. Today's quantum mechanics, photons and the idea of wave-particle duality bear only a minor resemblance to Newton's understanding of light.

In his *Hypothesis of Light* of 1675, Newton posited the existence of the ether to transmit forces between particles. In 1704, Newton published *Opticks*, in which he expounded his corpuscular theory of light. He considered light to be made up of extremely subtle corpuscles, that ordinary matter was made of grosser corpuscles and speculated that through a kind of alchemical transmutation, "Are not gross Bodies and Light convertible into one another, ...and may not Bodies receive much of their Activity from the Particles of Light which enter their Composition?" Newton also constructed a primitive form of a frictional electrostatic generator, using a glass globe.

The Beginnings of Diffractive Optics: The effects of diffraction of light were first carefully observed and characterised by Francesco Maria Grimaldi, who also coined the term *diffraction*, from the Latin *diffringere*, 'to break into pieces', referring to light breaking up into different directions. The results of Grimaldi's observations were published posthumously in 1665. Isaac Newton studied these effects and attributed them to inflexion of light rays. James Gregory (1638-1675) observed the diffraction patterns caused by a bird feather, which was effectively the first diffraction grating. In 1803, Thomas Young did his famous experiment observing interference from two closely spaced slits. Explaining his results by interference of the waves emanating from the two different slits, he deduced that light must propagate as waves. Augustin-Jean Fresnel did more definitive studies and calculations of diffraction, published in 1815 and 1818, and thereby gave great support to the wave theory of light that had been advanced by Christiaan Huygens and reinvigorated by Young, against Newton's particle theory.

Lenses and Lensmaking

The earliest known lenses were made from polished crystal, often quartz, and have been dated as early as 700 BC for Assyrian lenses such as the Layard / Nimrud lens. There are many similar lenses from ancient Egypt, Greece and Babylon. The ancient Romans and Greeks filled glass spheres with water to make lenses.

Glass lenses were not thought of until the Middle Ages. Ibn al-Haytham (Alhacen) wrote about the effects of pinhole and concave lenses in his *Book of Optics*, which was influential in the development of the modern telescope. The earliest evidence of "a magnifying device, a convex lens forming a magnified image," also dates back to his *Book of Optics*. Roger Bacon used parts of glass spheres as magnifying glasses and recommended them to be used to help people read. Roger Bacon got his inspiration from Alhacen in the 11th century. He discovered that light reflects from objects and does not get released from them. Around 1284, in Italy, Salvino D'Armate is credited with inventing the first wearable eye glasses.

Between the 11th and 13th century, the "reading stones" were invented. Often used by monks to assist in illuminating manuscripts, these were primitive plano-convex lenses initially made by cutting a glass sphere in half. As the stones were experimented with, it was slowly understood that shallower lenses magnified more effectively.

There is some documentary evidence, but no surviving designs or physical evidence, that the principles of telescopes were known in the late-16th century. Leonard Digges, Taqi al-Din and Giambattista della Porta independently developed rudimentary telescopes in the 1570s and 1580s. However, the earliest known working telescopes were the refracting telescopes that appeared in the Netherlands in 1608. Their development is credited to three individuals: Hans Lippershey and Zacharias Janssen, who were spectacle makers in Middelburg, and Jacob Metius of Alkmaar. Galileo greatly improved upon these designs the following year. Niccolo Zucchi is credited with constructing the first reflecting telescope in 1616. In 1668, Isaac Newton designed an improved reflecting telescope that bears his name, the Newtonian reflector.

The first microscope was made around 1595 in Middleburg, Holland. Three different eyeglass makers have been given credit for the invention: Hans Lippershey (who also developed the first real telescope); Hans Janssen; and his son, Zacharias. The coining of the name "microscope" has been credited to Giovanni Faber, who gave that name to Galileo Galilei's compound microscope in 1625.

Quantum Optics

Light is made up of particles called *photons* and hence inherently is "grainy" (quantised). Quantum optics is the study of the nature and effects of light as quantised photons. The first indication that light might be quantised came from Max

Planck in 1899 when he correctly modelled blackbody radiation by assuming that the exchange of energy between light and matter only occurred in discrete amounts he called *quanta*. It was unknown whether the source of this discreteness was the matter or the light. In 1905, Albert Einstein published the theory of the photoelectric effect.

It appeared that the only possible explanation for the effect was the existence of particles of light called *photons*. Later, Niel Bohr showed that the atoms were also quantised, in the sense that they could only emit discrete amounts of energy. The understanding of the interaction between light and matter following from these developments not only formed the basis of quantum optics but also were crucial for the development of quantum mechanics as a whole. However, the subfields of quantum mechanics dealing with matter-light interaction were principally regarded as research into matter rather than into light and hence, one rather spoke of atom physics and quantum electronics.

This changed with the invention of the maser in 1953 and the laser in 1960. Laser science – i.e., research into principles, design and application of these devices – became an important field, and the quantum mechanics underlying the laser's principles was studied now with more emphasis on the properties of light, and the name *quantum optics* became customary.

As laser science needed good theoretical foundations, and also because research into these soon proved very fruitful, interest in quantum optics rose. Following the work of Dirac in quantum field theory, George Sudarshan, Roy J. Glauber, and Leonard Mandel applied quantum theory to the electromagnetic field in the 1950s and 1960s to gain a more detailed understanding of photodetection and the statistics of light. This led to the introduction of the coherent state as a quantum description of laser light and the realisation that some states of light could not be described with classical waves. In 1977, Kimble *et al.* demonstrated the first source of light which required a quantum description: a single atom that emitted one photon at a time. This was the first conclusive evidence that light was made up of photons. Another quantum state of light with certain advantages over any classical state, squeezed light, was soon proposed.

At the same time, development of short and ultrashort laser pulses – created by Q switching and modelocking techniques – opened the way to the study of unimaginably fast (ultrafast) processes. Applications for solid state research (e.g. Raman spectroscopy) were found, and mechanical forces of light on matter were studied. The latter led to levitating and positioning clouds of atoms or even small biological samples in an optical trap or optical tweezers by laser beam. This, along with Doppler cooling was the crucial technology needed to achieve the celebrated Bose-Einstein condensation.

Other remarkable results are the demonstration of quantum entanglement, quantum teleportation, and (recently, in 1995) quantum logic gates. The latter are of much interest in quantum information theory, a subject which partly emerged from quantum optics, partly from theoretical computer science.

Today's fields of interest among quantum optics researchers include parametric down-conversion, parametric oscillation, even shorter (attosecond) light pulses, use of quantum optics for quantum information, manipulation of single atoms, Bose-Einstein condensates, their application, and how to manipulate them (a sub-field often called *atom optics*), and much more. Research into quantum optics that aims to bring photons into use for information transfer and computation is now often called *photonics* to emphasise the claim that photons and photonics will take the role that electrons and electronics now have.

Theory of Relativity

Special Theory of Relativity by Einstein

A major turning point in the Field theories proposed from time to time to interpret the physical nature of the universe was the formulation of the special theory of relativity in 1905 by Albert Einstein (1879-1955). It has emerged as a consequence of the failure of the attempts to discover the existence of the enigmatical luminiferous ether which was coined out of necessity to explain the propagation of light waves by Huygen and was imagined to fill all space and to be a perfectly elastic solid.

Later in 1864 Maxwell gave his electromagnetic theory of radiation, which also postulated the existence of a medium endowed with the properties of permittivity and permeability and the same were considered as the attributes of Huyget's luminiferous ether. Assuming its existence therefore, it might be conjectured that the motion of the planets and stars would in some way be affected by such an all pervading medium through which they were moving, but astronomical observations, accumulated over centuries, indicated no such effect. Since all speculations about the existence of ether stem from its properties as a medium for the propagation of light waves, it was expected that a sensitive optical experiment to determine whether light waves could be influenced by the motion of the earth through the ether, might settle the issue. The object of such an experiment would be to discover the ether drift equivalent to the orbital velocity of the earth of 29.6 km/sec around the sun, assuming the ether to serve as a stationary frame of reference. It was to find the effects of this motion on the velocity of light that an experiment was designed by A. Michelson and H. Morley in 1887, which we shall now describe in detail.

The Michelson and Morley Experiment: The object of the experiment was to test whether there was any relative motion between the earth and the luminiferous ether. The apparatus utilised was essentially a Michelson interferometer with its arm OM_2 parallel to the velocity of the earth around the sun. To avoid distortion of the instrument by strain the entire apparatus was mounted on a rigid platform floating on mercury in an iron trough and capable of slow rotation about an axis perpendicular to the plane of paper at a rate of ten complete rotations per hour. A beam of light originating at S is partly reflected at the back surface of A to M_1 and partly transmitted to M_2; M_1 and M_2 being optically equidistant from O, a point in the rear surface of A. The two beams are reflected from the front surfaces of M_1 and M_2 and meet at O, where they produce an interference pattern consisting of dark and bright band, visible to an observer at E. Let the length of each optical path be d.

Suppose the speed of light in a stationary ether is c and that of earth round the sun is v. The speed of light along OM_2 relative to the apparatus is $(c - v)$ and on return from M_2 to O its speed is $(c + v)$. The total time t, to go from O to M_2 and back to O is thus

$$t_1 = \frac{d}{c-v} + \frac{d}{c+d}$$

$$= 2d.\frac{c}{c^2 - v^2}$$

$$= \frac{2d}{c}\left[1+\frac{v^2}{c^2}+\frac{v^4}{c^4}+...\right]$$

$$= \frac{2d}{c}+\frac{2dv^2}{c^2} \text{ approx.} \qquad ...(i)$$

using binomial theorem and neglecting powers of v/c higher than two.

In going from O to M_1 and back to O, the actual path of the light beam through the ether is along $OM_1'O'$. This is due to the motion of the apparatus through the ether. The speed of light along this path is c so that its speed along the path OM_1 and M_1O is $(c^2 - v^2)^{1/2}$. The time taken by the light to go from O to M_1 and back to O is given by

$$t_2 = \frac{2d}{\left(c^2 - v^2\right)^{1/2}}$$

$$= \frac{2d}{c}\left[1+\frac{v^2}{2c^2}+...\right] \text{ by binomial theorem}$$

$$= \frac{2d}{c}+\frac{dv^2}{c^3} \text{ approx.} \qquad ...(ii)$$

Hence, ΔT the difference in the times for the two paths is given by

$$\Delta T = t_1 - t_2 = \frac{2d}{c}+\frac{2dv^2}{c^3}-\frac{2d}{c}-\frac{dv^3}{c^3} = \frac{dv^3}{c^3}$$

There is therefore a path retardation of $\Delta T \times c = dv^2/c^2$ between the two paths.

After having noted the position of the interference fringes with OM_2 parallel to the motion of the earth the whole apparatus is rotated through 90° so that the path OM_1 becomes parallel to the direction of the earth's motion, *i.e., the two paths interchange.* Such a rotation makes the path OM_1 and back optically longer than OM_2 and back by dv^2/c^2 instead of being the shorter, thus introducing a total path retardation between the two interfering beams,

$$\delta = 2\Delta T \times c = \frac{2dv^2}{c^2}$$

Hence, a displacement of interference fringes will occur. If f be the fraction of the fringe shift, then

$$f\lambda = \frac{2dv^2}{c^2}$$

or

$$f = \frac{2d}{\lambda}\left(\frac{v}{c}\right)^2 \qquad ...(iii)$$

Now from the known motion of the earth $\left(\frac{v}{c}\right)^2 = 10^{-8}$ approx.; d was taken equal to 11 m and $\lambda = 6 \times 10^{-7}$ m therefore we have $f = \frac{2 \times 11}{6 \times 10^{-7}} \times 10^{-8} = 0.37$ or about $\frac{1}{3}$ of a fringe width.

The actual displacement observed was certainly less than 20th part of this amount and probably less than the fortieth part. *Thus a diligent search failed to measure ether drift.* This means that the relative velocity between the earth and the ether is zero. The experiment has been repeated and other experiments have also been performed. The weight of evidence is in favour of the view that the existence of the ether cannot be proved by such experiments.

Recently Dr. C.H. Townes of Columbia University (1958) used the *maser*, an instrument capable of measuring very short time intervals with very great accuracy to recheck the Michelson-Morley experiment. The results are in close agreement with that of Michelson and Morley, that is, the speed of light in empty space is constant and is not affected by the motion of the observer or the source through the ether. *We, therefore, conclude that ether does not exist.* This led Einstein to re-examine the fundamental laws of mechanics and he then proposed his theory of relativity which affords a satisfactory explanation of the experiment besides having the advantage of logical perfection and security of foundations, which are the essential attributes of a principal theory. Before we take up the postulates of the Einstein's Special Theory of Relativity and its consequences, let us briefly explain Newtonian relativity of inertial systems. (*An inertial system of reference in physics is one to which events are spatially referred and in which, when there are no forces, an object moves with constant rectilinear velocity).*

Relativity Theory of Newton

The student is familiar with Newton's three laws of motion. These laws hold good to a fair degree of approximation when referred to a system of axes relative to earth. Let us now investigate how these laws are modified in a frame of reference moving with a velocity v in the positive X-direction, say.

Let F $(OXYZ)$ be a fixed frame of reference and F' $(O'X'Y'Z')$ a frame moving with uniform velocity v in the positive X-direction so that O, O' coincide at time $t = 0$. Let each describe the rectangular coordinates of the same point P. According to an observer in frame F the coordinates of P at time t are x, y, z, and according to that in frame F' they are y', y', z'.

The necessary equations of transformation for the frame F, in motion relative to the fixed frame F are

$$x' = x - vt;$$

$$y' = y; \ z' = z \qquad \text{... (1)}$$

Differentiating these equations with respect to time, we get

$$\dot{x}' = \dot{x} - v \,; \dot{y}' = \dot{y}' \,; \dot{z}' = \dot{z} \qquad \text{...(2)}$$

This is described as the *classical transformation for non-accelerated systems.* When critically examined it implies that space and time are absolute entities independent of each other and independent of the frame. These equations refer to the components of the velocity of the particle as determined by observers in the two coordinate frames.

The corresponding equations of acceleration in the two frames are obtained by a second differentiation and are

$$\ddot{x}' = \ddot{x} ; \ddot{y}' ; \ddot{z}' = \ddot{z} \quad \text{... (3)}$$

It follows that the acceleration of the particle is the same in the two frames and hence Newton's second law of motion $F=ma$, is equally valid when transformed to a set of axes moving with uniform velocity with respect to the original one or in other words, the Newton's laws of motion are *invariant, i.e. unaltered in form by a change of axes,* under the so-called Galilean-Newtonian type of transformation.

This obviously leads to the conclusion that an observer within an inertial moving system cannot detect his motion by any dynamical experiments performed wholly within that moving system. This result is called Newtonian or classical relativity and classical dynamics confirms it. Further, it may be noted that 'classical dynamics' search [Michelson-Morley experiment] for a fixed frame of reference in stationary ether has failed. The wave theory of light and Maxwell's electrodynamics sought luminiferous ether as a unique reference frame but all attempts to establish that frame were fruitless. Both experiment and theory show that if electromagnetic measurements stated with respect to fixed axes are transformed to *axes moving with velocity v in the X-direction,* the form of transformation rule changes. In order to maintain *a principle of relativity* it is necessary to be able to transform from axes at rest to axes in motion *without changing the forms of the equations of motion.* That is, Maxwell's equations of electromagnetic field should remain *invariant.* But this is not so. All these failures demanded a new orientation of the relativity which would include dynamical as well as electromagnetic phenomena.

Fundamental Postulates of Einstein's Special Theory of Relativity: In attempting to explain the results of Michelson-Morley experiment and other similar experiments to detect motion through the ether, Einstein in 1905 arrived at a new idea of assuming an ether filling empty space as a *meaningless concept* and that only motion *relative* to material bodies has physical significance. Accordingly with a view to harmonise this assumption with the known laws of optics and electromagnetism, he based his new theory which is known as special or restricted theory of relativity on two fundamental postulates:

(1) The *first postulate* is based on classical theory and States that 'the fundamental physical laws should have the same mathematical form in all inertial systems in uniform translatory motion with respect to one another.' This is another way of stating the requirement that the equations of a physical theory be *invariant* with respect to coordinate systems moving with different velocities. This is the *principle of equivalence.*

(2) The *second postulate* is based on the results of Michelson-Morley experiment and other optical and electrodynamic measurements. It may be stated as follows:

The speed of light in empty space is constant and is independent of the source and the observer. No signal or energy can be transmitted with a speed greater than the speed of light.

This means that a light wave leaving a star will have a constant velocity of 3×10^5 km per second relative to an observer regardless of the fact whether he and the star are approaching or receding from each other with any velocity, whatsoever.

The speed of light is defined as the distance traversed by the beam of light divided by the time required to traverse this distance as measured by any observer in a given reference frame.

Transformation of Lorentz-Einstein

If the postulates of Einstein's relativity are accepted we shall have to formulate equations of transformation expressing the relation between measurements made by observers in different frames of reference so as to enable us to transform the coordinates of one frame into those of other. Let us consider again a reference frame F' ($O'X'Y'Z'$) moving relative to frame $F(OXYZ)$ with uniform velocity v in the positive X-direction.

Suppose an event observed in the F frame occurs at coordinates x, y, z at time t, the same event observed in the F' frame has coordinates x', y', z' and will be observed at a time t', the two instants t and t' must not necessarily be the same even though the clocks used by the two observers stationed in them are identical in all respects and have been properly synchronised at the instant they passed each other at the origin of coordinates. Thus to characterise an event we require in addition to the three space coordinates x, y, z, the time t as the *fourth* coordinate. This mean that space and time are not separate entities but are inextricably bound to each other forming a four dimensional space-time continuum so that one without the other becomes meaningless.

To derive the equations of transformation of space and time variables appropriate to the special theory of relativity we choose a physical event, say, the emission of a pulse of light from a source and consider the description of this event by the two observers subject to two important limitations: i) The measuring instruments in the two frames cannot communicate directly except when they coincide in position. ii) The only means of communication between the two frames is a pulse of light. Suppose the source of light is at O, the origin of coordinates of frame F and suppose that O' the origin of coordinates of frame F' coincides with O at the instant the pulse of light is emitted and let $t = t' = 0$ at this instant. According to second postulate of relativity the speed of light is constant and independent of the motion of the observer, hence each one will note a spherical wave spreading out from his origin of coordinates. If c is the speed of light, the position of the spherical wave will be given by

$$x^2 + y^2 + z^2 = c^2t^2 \qquad ...(4)$$

in the F frame,

and by $x'^2 + y'^2 + z'^2 = c^2t'^2$ ...(5)

in the F' frame.

Thus, $\dfrac{x^2+y^2+z^2}{t^2} = \dfrac{x'^3+y'^2+z'^2}{t'^2} = c^2$ ..(6)

(constant) for either frame.

To find the relation between x and x', t and t' we use the Galilean transformation as a guide and assume that the equations will be *linear* and of the forms,

$$x' = K(x - vt) \quad ...(a)$$

$$y' = y \quad ...(b) \qquad ...(7)$$

$$z' = z \quad ...(c)$$

$$t' = At + Bx \quad ...(d)$$

where the constants, K, A and B are to be determined.

Substituting set of equations (7) in equation (5) we have

$K^2 (x - vt)^2 + y^2 + z^2 = c^2 (At + Bx)^2$

Expanding the above equation and collecting terms gives

$(K^2 - B^2c^2)x^2 + y^2 + z^2 - (A^3C^2 - K^2v^2)t^2 - 2(ABc^2 + K^2v)\, xt = 0$

Now $x^2 + y^2 + z^2 - c^2 t^2 = 0$ by equation (4)

Comparing coefficients of like terms in them, we get

$$K^2 - B^2C^2 = 1 \quad ...(i)$$

$$A^2C^2 - K^2v^2 = c^2 \quad ...(ii)$$

$$ABC^2 - K^2v = 0 \quad ...(iii)$$

The solution of these equations yields

$$K = A = \left(a - \frac{v^2}{c^2}\right)^{-\frac{1}{2}}$$

and $$B = \frac{Kv}{v^2},$$

so that the equations of transformation become

$$x' = K(x - vt) \quad ...(a)$$

$$y' = y \quad ...(b)$$

$$z' = z \quad ...(c)$$

$$t' = K\left(t - \frac{vx}{c^2}\right) \quad ...(d) \qquad ... (8)$$

$$K = \left(1 - \frac{v^2}{c^2}\right)^{-\frac{1}{2}} \quad ...(e)$$

·quations are known as Lorentz-Einstein equations of .mation of space and time coordinates and enable us to transform the coordinates of the moving frame $(x'_{,} y', z', t')$ into those of the stationary frame (x, y, z, t).

The transformation which is inverse of (8) can be obtained by replacing v by $-v$.

$$x' = K(x + vt) \quad ...(a)$$

$$y' = y \quad ...(b)$$

$$z' = z \quad ...(c) \qquad ...(9)$$

$$t' = K\left(t' + \frac{vx'}{c^2}\right) \quad ...(d)$$

This follows immediately from the fact that the frame F is moving with a velocity $-v$ relative to frame F'. The value of K however remains unchanged. It may be noted that in both these sets of equations the lengths perpendicular to the direction of motion are not affected. If the motion of the moving frame were not along the x-axis exclusively, then y', z', coordinates would also have transformation equations similar to those for x'.

Lorentz found that the results for a system moving with velocity v in the X-direction are simplified considerably when they were expressed as a function of $\left(t - \frac{vx}{c^2}\right)$ as derived above instead of as a function of t. Hence, $\left(t - \frac{vx}{c^2}\right)$ may be regarded as a sort of local time for the plane specified by X. According to the *principle of relatively* then the earth's motion has no influence, whatsoever, upon optical phenomena, because each observer carries with him his own system of space and time and the units which he employs to measure space and time adjust automatically as to eliminate the influence of motion.

Maxwell's equations of electromagnetic field are invariant with respect to this transformation. They have the same form as for a system at rest, that it, if equations are expressed in terms of x, y, z, t and x', y', z', t' are substituted by means of the formulae above, the form of the equations is unaltered, since Maxwell's equations sum up all optical and electrical phenomena. All physical phenomena are supposed to be essentially electrical and they are hence unaffected by motion.

The distinctive feature of the new hypothesis is that it is quantitative and Einstein showed that it is necessary. There are certain phenomena, which cannot be explained on the assumption that space and time are absolute and hence that assumption must go. Relativity consequently is now accepted as a faith. It is inadvisable to devote attention to its paradoxical aspects.

Some Consequences of the Special Theory: Let us now see how the fundamental concepts of length, time and mass are modified as a consequence of Lorentz-Einstein equations of transformation.

(i) *Relativity of Length (Contraction of Space):* From Lorentz-Einstein equations of transformation it can be shown that length or space is not absolute but *relative*. Suppose a rigid rod of length L_2 is lying in the frame F' and the axis x' is taken parallel to the length of the rod. Then in that frame the length of the rod can be expressed as

$$L_0 = x'_2 - x'_1, \qquad ...(10)$$

where x_1' and x_2' are the coordinates of the two ends of the rod.

The length of the rod in the moving frame F' as measured by an observer in the fixed frame F would be the distance between the coordinates x_1 and x_2 of the ends of the rod at a given time t, so that

$$L = x_2 - x_1 \qquad ...(11)$$

Using equation 8(a) we find that

$$\begin{aligned} L_0 &= x_2' - x_1' \\ &= K(x_2 - vt) - K(x_1 - vt) \\ &= K(x_2 - x_1) = KL \end{aligned}$$

from which

$$L = \frac{L_o}{K} = L_0 \left(1 - \frac{v^2}{c^2}\right)^{\frac{1}{2}} \qquad ...(12)$$

Thus the measured value of the length of the rod will be less when it is moving parallel to its length than when it is at rest with respect to an observer, the ratio of shortening being $\sqrt{1 - v^2 / c^2}$. If v has the value c, *i.e.*, if the matter moves with the velocity of lights *its length becomes zero*. This is clearly not possible, hence a particle can approach but never attains the velocity of light. The conctraction of an equivalent magnitude when physical objects move through the ether was originally postulated by Lorentz-Fitzgerald contraction hypothesis based upon electromagnetic conceptions which were considered justified because of the electrical theory of matter. It has recently (1959) been shown by Terrell that there is no substance in the contraction hypothesis. *It is only an apparent contraction and is inherent in the process of measurement thus accounting for the negative result of Michelson-Morley experiment.*

The apparent distortion of a moving object when seen by a stationary observer is due to the fact that the light forming the image is emitted at different times and from different positions of the object in space. This is just sufficient to cancel the Lorentz contraction. It may further be noted that in the derivation of equation (12) it is assumed that the light from the two ends of the rod is emitted and recorded simultaneously by the measuring instruments.

(ii) *Relativity of Time (Dilation of Time):* Let us now determine the relative values of time intervals as measured by the observers in two frames of reference in uniform relative motion. The equations 8(*d*) and 9(*d*),

$$t' = K\left(t - \frac{vx}{c^2}\right)$$

and $$t = K\left(t' + \frac{vx'}{c^2}\right), \text{ respectively}$$

clearly show that the measurement of time and time intervals by two different observers in uniform relative motion with respect to one another will depend upon this motion even though the same or identical clocks are used in their measurements.

Clocks moving with respect to an observer appear to go slow than they do when at rest with respect to him. If an observer in the frame F observes by his clock time interval Δt some event requires in the frame F', then his time interval will be longer than $\Delta t'$ determined by a clock in the moving frame F'. This effect is called *time dilation.*

To understand how time dilation comes about let us consider a clock at the point x' in the moving frame F'. When an observer in frame F' finds that time is t'_1, one in frame F will find it t_1, whence from equation 9 (*d*)

$$t_1 = K\left(t'_1 + \frac{vx'}{c^2}\right)$$

After a time interval $\Delta t'$ (to him), the observer in the moving frame finds that the time is now t_2' according to his clock such that

$$\Delta t' = t'_2 - t'_1 \qquad \text{...(13)}$$

The observer in F, however, measures the end of the same time interval to be

$$t_2 = K\left(t_2' + \frac{vx'}{c^2}\right)$$

So to him the duration of time interval Δt is

$$\Delta t = t_2 - t_1$$

$$= K\,(t_2' - t_1')'$$

$$= K\,\Delta t'$$

to that $$\Delta t > \Delta t'. \qquad ...(14)$$

since $$K = \left(1 - \frac{v^2}{c^2}\right)^{-1/2}$$

A stationary clock thus measures a longer time interval between events occurring in a moving frame of reference than does a clock in the moving frame.

In other words, a clock moving with velocity v relative to an observer in frame F will lag behind one at rest in F' in the ratio $\sqrt{1 - v^2 / c^2} : 1$. It always appears to go slow to an observer in F than it does to an observer at rest in its own frame, *viz.*, the time processes are slowed down resulting in the dilation of time. *This opinion is reciprocal* for a clock at the origin of the frame F' where it obeys the relationship $\Delta t' = K\Delta t$ leading to a similar conclusion. This may be difficult to understand in an intuitive way; it may take some time before you grasp the exact implication of this concept. The root of apparent paradox is the *invariance* of c. We see from the equations that if the moving frame is proceeding with the velocity of light that is $v = c$, then any time interval $\Delta t'$ occurring in it will be measured as Δt = infinity in the stationary frame-that is, the moving clock will appear to have stopped.

The effect must hold for any type of clock. In particular if T be the decay half-life of mesons or radioactive matter as measured in the frame F in which the particles are at rest, then

$T' = T/\left(1 - \frac{v^2}{c^2}\right)^{\frac{1}{2}}$ is the decay half-life observed in a frame F' in which the particles are moving with velocity v. This predicted time has been confirmed by the experiments on the measurement of life-time of π mesons carried out by Durbin, Loar and Havens (1952). The π^+ mesons in a frame in which they are at rest have a mean life before decaying of about 2.5×10^{-2} sec. If the relativistic time dilation effect did not exist then assuming their velocity nearly equal to c, they would traverse a mean distance equal to $2.5 \times 10^{-8} \times 3 \times 10^8$ m/ sec = 75 m, before decaying. Actually they have been found to travel much farther than this because of time dilation, traversing sometimes a distance about 100 times the distance they would traverse before decay without the time dilation effect. The experiments on high energy particles generated in nuclear disintegration processes provide the physicist with tests for special relativity everyday.

We shall illustrate the time dilation of equation (14) by an example pertaining to the decay of μ mesons.

Example: *The mean life of a muon at rest is about 2×10^{-6} s. If muons in cosmic rays have an average speed of 0.998 c, find the mean life of these muons and approximate distance they travel before decaying.*

Let y_0 be the altitude from ground at which a μ meson is produced. Its mean life $\Delta t'$ which is 2×10^{-6} seconds in its own *reference frame* will be extended in *our reference frame* on account of relative motion, to the value

$$\Delta t = K\Delta t'$$

$$= \frac{\Delta t'}{\sqrt{1 - v^2/c^2}}$$

$$= \frac{2 \times 10^{-5}\ \text{s}}{\sqrt{1-(0.998\, c/c)^2}}$$

$$= \frac{2 \times 10^{-5}}{0.063} \text{ s}$$

$$= 32 \times 10^{-6} \text{ s}$$

almost 16 times greater than when it is at rest with respect to us. In 32×10^{-6} second a meson whose average speed is 0.998 c can travel a distance

$$y_0 = vt$$

$$= 0.998 \times 3 \times 10^8 \text{ ms}^{-1} \times 32 \times 10^{-6} \text{ s}$$

$$= 9500 \text{ m}$$

$$= 9.5 \text{ km.}$$

This explains how, despite their brief life spans, it is possible for the mesons to reach the ground from the high altitudes at which they are actually produced.

On the other hand, in classical physics time flows on *uniformly* for all observers on the same frame or on different frames relatively in motion without regard to anything external. Thus there was supposed to be a single time scale valid everywhere, whereas in relativity each observer has its own local time, which may be quite different from the times of other observers, *i.e.*, $t \neq t'$. The Newtonian conception of *'absolute'* time fails and the notion of *'simultaneity of events'* in respect of two frames relatively in uniform motion becomes untenable.

(iii) *Relativity of Simultaneity*: From equations (8) and (9) we see that two events which are simultaneous in the frame F but occur at different places are not simultaneous in the frame F'.

For if $t_1 = t_2$

$$t_1' = K\left(t_1 - \frac{v}{c_2} x_1\right)$$

and

$$t_2' = K\left(t_1 - \frac{v}{c_2} x_2\right) \quad ...(15)$$

where $K = \left(1 - \frac{v^2}{c^2}\right)^{-1/2}$

Thus t_1' and t_2' are not equal if $x_1 \neq x_2$.

We shall Illustrate the relativity of simultaneity by an idealised. Suppose a fast moving train 21.6×10^5 km long is travelling in a straight line with a uniform speed of 2.4×10^5 km/sec past a stationary observer *O* beside the railway track.

The train has two doors *A* and *B* one at either end which open on the arrival of a light flash. It is so arranged that they are visible to an observer *P* seated at the mid-point of the train and also to the stationary observer *O*, *P* sends a light flash just when he is opposite *O* and sees that both doors *A* and *B* open simultaneously

$$\frac{10.8 \times 10^5 \text{ km}}{3 \times 19^5 \text{ km / sec}} = 3.6 \text{ sec}$$

after the flash was sent (3×10^6 km per sec. is the speed of light). This is in accord with Michelson's celebrated experiment regarding the constancy of the speed of light.

The stationary observer *O* however, sees that the rear door B opens,

$$\frac{10.8 \times 10^5\ km}{(3 \times 2.4) \times 10^5 km/sec} = 2 \text{ secs}$$

after the flash was sent, since the rear of the train moves at 2.4×10^5 km per second to meet the flash travelling at 3×10^6 km/ sec. But the front door *A* is moving forwards at 2.4×10^5 km/sec and the light flash travelling at 3×10^5 km/sec has to catch it up. It will, therefore, take

$$\frac{10.8 \times 10^5 \text{ km}}{(3 - 2.4) \times 10^5 \text{km / sec}} = 18 \text{ secs}$$

for the front door *A* to open, that is, for the stationary observer *O*, the front door *A* will open 16 sec later than the other door *B* opens.

Thus the two identical events occur simultaneously for P but at an interval of 16 secs for O. The relativity of simultaneity thus abolishes the idea of "absolute time", *i.e.*, time lapses at the same rate everywhere in the universe. In other words, the moment of occurrence of an event anywhere in the universe was the same everywhere else, however, the relativity of simultaneity has established that there is no such thing as same time for events as observed by observers in different frames of reference.

Note: In view of contraction of the speeding train the actual interval between the opening of the rear and front doors according to observer O will be 9.6 seconds and not 16 seconds, calculate how ?

(iv) *Relativity of Velocity:* The Lorentz transformation leads to important formulas for the transformation of velocities from one frame to another. Suppose a particle is moving with a velocity w in the frame F' parallel to the *positive* direction of X -axis in the $X'Y'$ plane. According to classical relativity the velocity u_m of this particle relative to the stationary frame F is given by the well known equation

$$u_x = v + w, \qquad \text{...(16)}$$

where v is the uniform velocity of F' relative to F. We will now see that this relation does not hold when v is comparable to c.

The position of the particle at any time t' in the frame F' is given by the usual equation

$$x' = wt' \qquad \text{...(17)}$$

assuming that the particle starts at the Y'-axis at time $t' = 0$. Now the Lorentz-Einstein equations of transformation give

$$x' = K(x - vt) \qquad \textit{cf. } \text{equation } 8(a)$$

$$t' = K\left(t - \frac{vx}{c^2}\right) \qquad \textit{cf. } \text{equation } 8(d)$$

Substituting the values of x' and t' in equation (17), we have

$$K(x - vt) = wK\left(t - \frac{vx}{c^2}\right)$$

from which

$$x = vt + wt - \frac{wvx}{c^2}$$

or $$x + \frac{wvx}{c^2} = (v + w)t$$

or $$x\left[1 + \frac{wv}{c^2}\right] = (v + w)t$$

so that $$x = \frac{v + w}{\left(1 + \frac{wv}{c^2}\right)}t$$

therefore $$u_x = \frac{dx}{dt}$$

$$= \frac{v + w}{\left(1 + \frac{wv}{c^2}\right)} \quad ...[18(a)]$$

If w is towards –ve X-direction

$$u_x = \frac{v - w}{\left(1 - \frac{wv}{c^2}\right)} \quad ...[18(b)]$$

Also $$u_y = 0$$

and $$u_z = 0$$

Equations 18 (*a*) and 18 (*b*) give the relativistic rule for transforming parallel velocities.

Or more generally,

If w be the velocity in any random direction in $X'Y'Z'$ frame, then the relativistic velocities in both Y' and Z' directions will also be affected and we can deduce that

$$u_y = \frac{y}{t} = \frac{y'}{K\left(t' + \frac{vx'}{c^2}\right)}$$ *cf.* equations [9(*b*) and (*d*)]

$$= \frac{y'/t'}{K\left(1 + \frac{v}{c^2} \cdot \frac{x'}{t'}\right)} = \frac{w_y}{K\left(1 + \frac{vwx_\infty}{c^2}\right)}$$

$$= \frac{w_y\left(1 - \frac{y^2}{c^2}\right)^{\frac{1}{2}}}{1 + \left(\frac{vw_\infty}{c^2}\right)} \qquad ...[18\ (c)]$$

Similarly,

$$u_z = \frac{w_z\left(1 - \frac{v^2}{c^2}\right)^{\frac{1}{2}}}{\left(1 + \frac{vw_z}{c^2}\right)} \qquad ...[18(d)$$

and from equation [18(*a*)] $u_x = \dfrac{v + w_x}{\left(1 + \frac{vw_x}{c^2}\right)}$, where $w_{x'}$, $w_{y'}$, $w_{z'}$, are, respectively components in X', Y' and Z' directions in the frame F'.

Also note that equations 18 (*a*) and 18 (*b*) reduce to the Newtonian form $u_{x=} v \pm w$, when v and w are small as compared with c in which case wv/c^2 is negligible in comparison with unity.

In the special case in which $w = \pm c$, we have

$$u_x = \frac{v \pm c}{\left(1 \pm \frac{vc}{c^2}\right)}$$

so that $u_x = \pm c.$ ...(19)

That is, *the observer in F also measures the velocity of the particle as the velocity of light.*

This result should not surprise us as it confirms the second postulate of Einstein's relativity that the velocity of light is constant and is not affected by the motion of the source and the observer.

It is clear that *maximum velocity ever attainable is that of light,* for the addition of the velocity v to a particle with the velocity of light c results in a final velocity c. This limit to the velocity attainable in nature is of fundamental importance.

An Important Problem: Consider a particle moving with constant velocity along x-axis in frame F. Let us find its velocity from the point of view of two observers in the frames F and F' assuming that F is stationary and F' is moving relatively to it with a velocity v along the x-axis. According to observer in F' the particle moves from the point x_1' at time t_1' to the point x_2' at time t_2'. He then ascribes to it a velocity

$$v_1' = (x_2' - x_1')/(t_2' - t_1')$$

In a similar manner, the velocity of the same particle is estimated by the observer in F to be

$$v_1' = (x_2 - x_1)/(t_2 - t_1)$$

Substituting from the transformation equation (9), for x_1, x_2, t_1, t_2 in the latter expression (remembering that v is the *relative* velocity of the two observers) leads to

$$v_1 = \frac{K\left(x_2' + vt_2'\right) - K\left(x_1' + vt_1'\right)}{K\left(t_2' + vx_2'/c^2\right) - K\left(t_1' + vx_1'/c^2\right)}$$

Dividing throughout by K and then rearranging gives

$$v_1 = \frac{\left(x_2' - x_1'\right) + \left(t_2' - t_1'\right)v}{\left(t_2' - t_1'\right) + v\left(x_2' - x_1'\right)/c^2}$$

Dividing top and bottom by $(t_2' - t_1')$ yields

$$v_1 = \frac{\left[\left(x_2' - x_1'\right)/\left(t_2' - t_1'\right)\right] + v}{1 + \left\{v/c^2\right\}\left\{\left(x_2' - x_1'\right)/\left(t_2' - t_1'\right)\right\}}$$

or $$v_1 = \frac{v + v_1'}{1 + vv_1' / c^2} \qquad ...(20)$$

This is the *law of addition of velocities,* and may be compared with the classical or Newtonian law, which states simply that if reference frames have a relative velocity v and if particle velocity estimated by F' is v_1' then that observed by F is just $v_1 = (v + v_1')$ since the denominator of the relativistic expression above is more than one the velocity measure in the stationary frame is less than $(v + v_1')$. So as observed from rest, moving objects in a moving frame are relatively slowed.

If the observer in the frame F' were to observe the phenomenon in the frame F and wishes to express v' in terms of *his* position and time system, he would have to use the transformation equation (8). If the substitutions are made, the result is, as the reader may verify,

$$v_1' = (v - v) / (1 - vv_1/c^2). \qquad ...(21)$$

Example: *Suppose two electrons in a stationary frame F move off with equal speeds of magnitude 0.9 c, one towards –X and other towards +X. What is the speed of one electron relative to the other?*

According to Galilean transformation their speed relative to each other measured in frame F is $0.9c - (-0.9c) = 1.8c$. This exceeds c.

But if we take $w = -0.9c$ so that frame F' keeps up with the electron going towards –X, then by equation 18 (*b*), the velocity of second electron relative to first, measured now in F', is

$$\frac{[0.9c - (-0.9c)]}{(1 + 0.9^2\,c^2 / c^2)} = \frac{1.8}{1.81}\,\text{c} = 0.9994\,\text{c}$$

which is a little less than c.

Example: *The components of the velocity of a particle in a frame F' are $x_a = x_y = 0.6c$. This is observed from a reference frame F which* *moves along the X-axis with speed – 0.8c. Derive the velocity components of the particle measured from frame F.*

Using equation 18 (*b*) we have

$$u_x = \frac{0.6c-(-0.8c)}{1+\frac{0.6\times 0.8c^2}{c^2}} = \frac{1.4}{1.48}\text{ c} = 0.94\text{ c}$$

$$u_y = \frac{w_y\left(1-\frac{v^2}{c^2}\right)^{1/2}}{1+\frac{vw_x}{c^2}} = \frac{0{,}6c\,(1-0.64)^{1/2}}{1+\frac{0.6\times 0.8\,c^2}{c^2}}$$

$$= \frac{0.6\times 0.6}{1.48}\,c = 0.24\text{ c}$$

and $u_z = 0.$

Note that although the relative motion of the two frames is parallel to X-axis, the velocity along *Y* direction is also affected. The slope of the path in AT plane is 1.0 in frame *F'* but becomes 0.24/94=0.25 in frame *F*.

Example: *A velocity vector has components* $u_x = dx/dt$, $u_y = dy\,/\,dt$ *and* $u_z = dz/dt$ *in the S-system, and* $u_x' = d_x' = d_x'\,/dt$, $u_y = d_y'\,/\,dt'$ *and* $u_z = dz'/dt'$ in the moving S'-system. Show that

$$u'_x = (u_x - v)\,/\left(1-\frac{vu_x}{c^2}\right)$$

$$u_y' = u_y\left(1-\frac{v^2}{c^2}\right)^{1/2}\Big/\left(1-\frac{vu_x}{c^2}\right)$$

and $$u_z' = u_z\left(1-\frac{v^2}{c^2}\right)^{1/2}\Big/\left(1-\frac{vu_x}{c^2}\right)$$

Find the Equations for the Inverse Transformation:

Putting equation (8) in the differential form

$$dx' = K\,(dx - vdt);\ dy; = dy;\ dz' = dz$$

and $$dt' = K\,(dt - vdx/c^2)$$

From these expressions by substitution, we have

$$u_x' = \frac{dx'}{dt'} = \frac{K(dx - vdt)}{K\left(dt - \dfrac{vdx}{c^2}\right)} - \frac{u_x - v}{1 - \dfrac{vu_x}{c^2}}$$

$$u_y' = \frac{dy'}{dt'} = \frac{dy}{K\left(d' - \dfrac{vdx}{c^2}\right)} = \frac{dy / dt}{K\left(1 - \dfrac{vdx}{c^2\, dt}\right)}$$

$$= \frac{dy}{K\left(1 - \dfrac{vux}{c^2}\right)} = \frac{u_y\left(1 - v^2 / c^2\right)^{1/2}}{1 - \dfrac{vu_x}{c^2}}$$

Similarly,

$$u_y' = \frac{dz'}{dt'} = \frac{u_z\left(1 - v^2 / c^2\right)^{1/2}}{1 - \dfrac{vu_x}{c^2}}$$

The inverse equations can be obtained by using equation (9) in differential form.

(v) ***Variation of Mass with Velocity:*** Another important consequence of the special theory of relativity is that the masses of a particle varies with velocity and is given by

$$m = Km_0 = \frac{m_0}{\left(1 - \dfrac{v^2}{c^2}\right)^{1/2}} \qquad ...(22)$$

in which m_0 is the mass of the particle when at rest with respect to an observer and m is the mass when it is moving with velocity v relative to it. This is in contradiction to the Newtonian theory of invariability of mass, *i.e.*, the mass of a body does not depend upon its velocity. Since the factor $\sqrt{1 - v^2 / c^2}$ in equation (22) is always less than unity, we find that mass increases with velocity. The increase in mass is about $\frac{1}{2}$ per cent upto a velocity of 1/10th that of light but rises much more rapidly with greater velocities being infinite when the velocity of light is reached. This relation has been experimentally confirmed by measurement of the masses of high speed electrons and β rays emitted by some radioactive substances by Bucherer and several other experimenters. The change of mass with velocity is very important in spectroscopic theory, for in the case of elliptic orbits the velocities of the electrons are so great when close to the nucleus that the mass change with velocity affects the energies of orbits considerably.

This fundamental result can be derived theoretically in several different ways, one of them being that in which we assume the principle of 'conservation of momentum' which is valid in all inertial frames of reference. But here we shall content ourselves by giving the expression only.

Relativistically therefore, we shall define the momentum p of a particle as $p = mv$, when m is the relativistic mass. Accordingly

$$P = \frac{m_0 v}{\sqrt{1 - v^2 / c^2}} \qquad \text{...(23)}$$

since $$m = \frac{m_0}{\sqrt{1 - v^2 / c^2}}$$

Table given below may be of interest to the student as it gives the value that the mass of an electron attains at different energies and velocities. The first and third columns, of course, would be the same for bodies of any *rest mass.*

Table showing relativistic mass increase of the electron

Velocity per cent C	*Energy × 10^{-4} ev:*	*Mass*
0	0	m_0
86	0.5	$2m_0$
94	1.0	$3m_0$
96	1.5	$4m_0$

Contd...

Velocity per cent C	*Energy* × 10^{-4} *ev:*	*Mass*
98	2.0	$5m_0$
99.5	4.5	$10m_0$
99.8	49.5	$100m_0$
100	∞	∞

The Relationship Between Mass and Energy of a Relativistic Particle: A particle of rest mass m_0 moving with a velocity v comparable with the velocity of light c is termed a '*relativistic particle.* In determining the kinetic energy of such a particle, we must take into account the mass variation with increase of velocity. This leads us to a modification of our ideas about energy. Suppose a force F acts parallel to the displacement ds of the particle, then the work done on it or the increase in kinetic energy is given by

$$dE_x = Fds$$

In classical physics force is defined as the time rate of change of momentum, so that

$$F = \frac{d}{dt}(mv) \quad \text{...(24)}$$

Therefore the change in kinetic energy,

$$dF_x = \left[\frac{d}{dt}(mv)\right] ds$$

$$= \frac{ds}{dt} . d(mv)$$

But since $\frac{ds}{dt} = v$, we can write this equation as

$$dE_k = vd\,(mv)$$

$$= v^2dm + mvdv,$$

since mass is also variable.

This expression can be simplified by using the relativity mass transformation equation (22) according to which

$$m = m_0 \Big/ \left(1 - \frac{v^2}{c^2}\right)^{1/2}.$$

which after squaring and rearranging becomes

$$m^2c^2 = m^2v^2 + m_0^2 c^2.$$

Differentiating this and noting that m_0 and c are constants we obtain

$$2mc^2dm = 3mv^2dm + 2m^2vdv$$

$$c^2dm = v^2dm + mvdv.$$

The right-hand side becomes identical with our expression for dE_k given above, so we have finally

$$dE_k = c^2dm. \qquad ...(25)$$

Thus a change in kinetic energy is *directly* proportional to the change in mass, the proportionality factor being the square of the velocity of light.

This is a very important relation in relativity indicating that a change in kinetic energy can be expressed in terms of mass as the variable. This equation is valid for change in energy in any form whatever.

Since the kinetic energy is zero when $v = 0$, then it is also zero when $m = m_0$, therefore integrating equation (25) we obtain

$$E_x = \int_0^{E_x} dE = c^2 \int_{m_0}^{m} dm = c^2(m - m_0) \quad ...(26)$$

Thus the kinetic energy of a moving body equals c^2 times its gain in mass due to the motion. This relation suggests that we may think of the increase in energy as the actual cause of the increase in mass and implies that in addition to the familiar forms of energy, such as kinetic and potential, etc., there is yet another kind 'mass energy'. According to this view equation (26) can be rewritten as

$$mc^2 = E_k + m_0 c^2. \qquad ...(26a)$$

The quantity m_0c^2 is called the *rest mass energy* of the particle. We can now define that total energy E of the particle as

$$E = mc^2 \quad \text{... (27)}$$

in which m is the relativistic mass of the particle. This equation leads to the idea of the *equivalence of mass and energy* and is certainly the most practical consequence of Einstein's theory of relativity.

Einstein's principle of relativistic mass energy equivalence may be stated as: "A mass m is equivalent to an amount of energy E and the equation relating these quantities is

$$E = mc^2,$$

where c is the speed of light".

This relation has received abundant experimental verification. It can be derived directly from the definition of the kinetic energy E_k of a moving body as the work done in bringing it from rest to its state of motion

i.e. $$E_x = \int_0^x F dx, \quad \text{...(28)}$$

where F is the component of the applied force in the direction of the displacement dx and x is the distance over which the force acts. Using the relativistic form of the second law of motion.

$$F = \frac{d(mv)}{dt}$$

Equation (28) becomes

$$E_x = \int_0^x \frac{d(mv)}{dt} dx$$

$$= \int_0^{mv} v d(mv)$$

$$= \int_0^v v d\left(\frac{m_6 \, v}{\left(1-v^2/c^2\right)^{1/2}}\right) \quad \text{...(29)}$$

Since $$m = \frac{m_9}{\left(1-v^2/c^2\right)^{1/2}}$$ ...by Equation (22)

Integrating Eq. (29) by parts $\left(\int u dw = uw - \int w du\right)$

$$\therefore \qquad E_k = \frac{m_0 V^2}{\sqrt{1 - v^2/c^2}} - m_0 \int_0^v \frac{v dv}{\sqrt{1 - v^2/c^2}}$$

$$= \frac{m_0 v^2}{\sqrt{1 - v^2/c^2}} + m_0 \int_0^v \frac{-c^2 v dv}{+c^2\sqrt{1 - v^2/c^2}}$$

$$= \frac{m_0 v^2}{\sqrt{1 - v^2/c^2}} + m_0 c^2 \left[\sqrt{1 - v^2/c^2}\right]_0^v$$

$$= \frac{m_0 v^2}{\sqrt{1 - v^2/c^2}} + m_0 c^2 \left[\sqrt{1 - v^2/c^2} - 1\right]$$

$$= \frac{m_0 v^2}{\sqrt{1 - v^2/c^2}} + m_0 c^2 \sqrt{1 - v^2/c^2} - m_0 c^2$$

$$= \frac{m_0 v^2}{\sqrt{1 - v^2/c^2}} + m_0 c^2$$

$$= mc^2 - m_9 c^2$$

$$\text{or} \qquad mc^2 = E_k + m_0 c^2 \qquad \text{...(30)}$$

The quantity m_0c^2 is called the *rest-mass energy* of the particle, thus we can define the *total energy* of the particle as

$$E = mc^2 \qquad \text{...(31)}$$

when m in the relativistic mass of the particle.

This equation leads us to the idea of *equivalence of mass and energy* and is known as Einstein's mass-energy equation.

This relation can also be expressed as:

$$E = \frac{m_0 v^2}{\sqrt{1 - v^2/c^2}}. \qquad \text{...(32)}$$

A very useful equation describing a relativistic particle is the one between the momentum and the total energy E giving us the invariant $E^2 - p^2c^2$. By equations (32) and (23)

$$E^2 - p^2c^2 = \frac{m_0^2\, c^4}{1 - v^2/c^2} - \frac{m_0^2\, v^2\, c^2}{1 - v^2/c^2}$$

$$= m_0^2\, c^4 \qquad \text{(constant)}$$

Expressing E in terms of momentum, we get

$$E = \left(m_0^2\, c^4 + p^2\, c^2\right)^{1/2} \qquad ...(33)$$

This is the relativistic *energy momentum relation.*

Showers of Cosmic Ray

These showers from outer space provide an other illustration of conversion of matter into energy. According to Millikan they are produced in some parts of the space where the temperature may be so low that the hydrogen nuclei may coalesce to form helium nucleus with liberation of energy corresponding to the *packing effect* in the form of radiation in accordance with Einstein's mass energy relationship, $E = mc^2$, where E is the energy liberated by a mass m and c is the velocity of light in vacuum. When four hydrogen nuclei unite to form helium nucleus, there is a loss of atomic weight equal to

$$1.0078 \times 4 - 4.002 = 0.0292$$

This loss of at. wt. measures liberation of energy

$$= 0.0292 \times 931 \times 10^6 \text{ eV}$$

$$[\because 1 \text{ at. wt.} = 931 \times 10^6 \text{ eV}]$$

$$= 27.1852 \times 10^6 \text{ eV}$$

The magnitude of the energy of cosmic rays gives likelihood to such a view.

Derivation of Classical Expression for E_k from Relativistic Expression: We can easily derive the classical expression and the kinetic energy from equation (26) which can also be written as

$$E_k = m_0c^2(K - 1), \qquad ...(34)$$

since $$m = km_0 \frac{m_0}{\left(1-\frac{v^2}{c^2}\right)^{\frac{1}{2}}}$$ *cf.* equation (22)

This relativistic expression for the kinetic energy of a particle will reduce to the more familiar classical expression when $v<c$. This can be seen by expanding the first term K in the bracket of equation (34) by binomial theorem.

Now $$K = \left(1-\frac{v^2}{c^2}\right)^{\frac{1}{2}} = 1 + \frac{1}{2}\frac{v^2}{c^2} + \ldots\ldots$$

neglecting terms in v/c above the second power.

Equation (34) thus becomes

$$E_k = m_0c^2\left[1+\frac{1}{2}\frac{v^2}{c^2}+\ldots-1\right]$$

$$= \frac{1}{2}\, m_0v^2.$$

This is the classical expression for the kinetic energy of a particle.

Thus we see that the relativity theory does not "overthrow" the classical theory but rather extends and modifies it. We can say that, in general, classical mechanics constitutes approximate form of mechanical theory that is *valid for* any motion which is slow as compared with the speed of light. The need for the relativity theory in atomic and nuclear physics comes in part from the fact that fundamental particles such as electrons, protons and mesons, etc. can travel with velocities approaching that of light with the result that kinematic and dynamical consequences of the Lorentz equations assume greater importance.

The Four Dimensional Continuum: The special theory of relativity dealing with systems moving relatively to each other with uniform velocity and its application to four dimensional space is the intermediate stage in the evolution of the General Theory of Relativity which deals with accelerated systems and gravitational fields. Thus general theory makes a number of predictions namely— (1) motion of the perihelion of mercury, (2) deflection of a ray of light and (3) shift of spectral lines in an intense gravitational field and so on, which have since been verified substantially. It also shows that Newton's law of gravitation is an approximation of a more general law enunciated by Einstein, according to which gravitational attraction was shown to be a property of space. Being beyond the scope of this book, the general theory will not be dealt with here. However, the concept of four dimensional continuous may be elaborated a bit.

When the equation of a line in cartesian coordinates is transformed we arrive at the following results:

The distance s between two points with coordinates x_1, y_1, z_1, and x_2, y_2, z_2 in the frame F, say, is given by

$$s^2 = (x_2 - x_1)^2 + (y_2 - y_1)^2 + (z_2 - z_1)^2 \quad ...(35)$$

For the second frame of reference F' moving with respect to the first with uniform velocity v, the application of the transformation already derived in equation (9), to the above equation yields

$$s^2 = K^2 \{(x_2' - x_1') + v(t_2' - t_1')\}^2 + (y_2' - y_1')^2 + (z_2' - z_1')^2 \quad ...(36)$$

This clearly depends upon v and is certainly not similar in form to the original equation. The equation (35) therefore is not *invariant* as required by the first postulate and hence not applicable to the space time continuum. If, however, this equation is replaced by

$$s^2 = (x_2 - x_1)^2 + (y_2 - y_1)^2 + (z_2 - z_1)^2 - c^2(t_2 - t_1) \quad ...(37)$$

it will be found by substituting the transformation values of equation (9), that it *does remain invariant* (verify).

Therefore, for a line element this invariant form of equation is

$$ds^2 = dx^2 + dy^2 + dz^2 - c^2dt^2 \quad ...(38)$$

If the velocity of light $c(=3 \times 10^8$ m/sec) is taken as unity for mathematical convenience, the invariant is slightly simplified to

$$ds^2 = dx^2 + dy^2 + dz^2 - dt^2 \quad \text{... (39)}$$

Physically interpreted, it means that a *length is not described by three dimensions dx, dy, dz* but in addition, *the time dimension dt* is forced upon us.

The quantity *ds* is called *point-event* and does not correspond exactly to a distance in space but is an *interval in a four dimensional continuum in which space and time are closely linked together.*

It will be noticed from the first form of the invariant for a line element, equation (38), that when $ds = 0$, we have

$$c^2 = (dx^2 + dy^2 + dz^2)/dt^2 \quad \text{...(40)}$$

i.e., something moving with the velocity of light. *This can only be a light photon* (or a ray of light), hence the equation $ds = 0$ is that for a ray of light.

In the four dimensional continuum the equivalent of a straight line is *a line making ds a minimum.* This is termed a *Geodesic.* The integral *ds, i.e.*, the expression $\int ds$, which denotes the *history* of a particle is called a *world line. The intersection of two world lines constitutes an ordinary 'event'.*

Some More Examples of Relativistic Calculations:

Example: *Find the energy produced by the complete annihilation of one gram of matter using the mass energy equation* $E = mc^2$.

If m is expressed in grams, E in ergs, then c must be expressed in cm per sec. Taking $c = 3 \times 10^{10}$ cm s^{-1} a mass of 1 g is equivalent to energy given by

$$E = 1\,g \times (3 \times 10^{10}\text{cms}^{-1})^2$$

$$= 9 \times 10^{20} \text{ ergs.}$$

$$= 9 \times 10^{18} \text{ joules} \qquad [10^7 \text{ ergs.}=1 \text{ Joule}]$$

Similarly, a mass of 1 kilogram is equivalent to

$$E = 1 \text{ kg} \times (3 \times 10^8 \text{ms}^{-1})^2$$

$$= 9 \times 10^{10} \text{ joules, or } 2.5 \times 10^{10} \text{ kW/hr.}$$

Note: The basic unit of nuclear physics is the *atomic mass unit* (*amu*), which is 13/12 of the rest mass of carbon 12.

One *amu* =1.66 × 10^{-27} kg and this is nearly the proton's mass. It is equivalent to

$$E = m_0c^2$$

$$= 1.66 \times 10^{-27} \text{ kg} \times (3 \times 10^8)^3 \text{ m}^2 \text{ s}^{-2}$$

$$= 1.49 \times 10^{-10} \text{ joule.}$$

In atomic and nuclear calculations, the erg or joule is an inconvenient unit of energy because the amounts of energy involved in single event usually are only small fractions of erg, say 10^{-4} to 10^{-12} erg. It is, therefore, the practice to express energies in units of electron-volts abbreviated as 'ev'. The electron volt is the energy acquired by any charged particle carrying a unit electronic charge, when it falls through a pot. diff. of 1 volt: it is equal to 1.6 × 10–12 erg or 1.6 × 10–12 joule. For conversion two other units are also used one equal to 10^3 ev called Kev and other equal to 10^6 ev called Mev.

1 *amu* may be expressed in ev and Mev.

$$1 \; amu = \frac{\text{energy in jounles}}{\text{joules per ev}}$$

$$= 1.49 \times 10^{-10} \text{ joule} \times \frac{1 \text{ ev}}{1.60 \times 10^{-19} \text{ joule}}$$

$$= 9.31 \times 10^8 \text{ ev} = 931 \text{ Mev.}$$

Example: *A relativistic particle whose rest mass is m_0 is moving with a velocity of 0.9 c. Determine (a) its relativistic mass (b) i_{ts} kinetic energy* (c) *why is it incorrect to use the expression $\frac{1}{2}mv^2$ for the kinetic energy of a relativistic particle ?*

The relativistic mass

$$m = \frac{m_0}{\left(1 - \frac{v^2}{c^2}\right)^{\frac{1}{2}}}$$

$$= \frac{m_0}{\left(1 - \frac{(0.9c)^2}{c^2}\right)^{\frac{1}{2}}} \quad \text{...}[\because v = 0.9c]$$

$$= \frac{m_0}{(0.9c)^{\frac{1}{2}}} = \frac{m}{0.436} = 2.3 \text{ m}_0$$

(1) The kinetic energy

$$E_k = mc^2 - m_0c^2$$
$$= 2.3\ m_0c^2 - m_0c^2$$
$$= 1.3\ \text{m}_0\text{c}^2$$

(*c*) Because we must take mass variation into account for a relativistic particle.

Example: *If an observer in motion moves away with a velocity c/2, how far will it be from the stationary observer in one second from the relativistic point of view ?*

According to classical mechanics in one second he will proceed a distance $c/2$. Relativistic equation 8(*a*) on the other hand gives

$$x_1' = K\,(x_1 - vt)$$

and $$x_2' = K\,(x_2 - vt)$$

$\therefore$ $$x_2' - x_1' = K\,(x_2 - x_1)$$

Let $X_2' - x_1'$ be the distance traversed in time t by the moving observer moving with a velocity $c/2$ as measured in its own frame, then

$$x_2' - x_1' = ct/2$$

$\therefore$ distance as measured from the point of view of the stationary observer

$$x_2 - x_1 = \frac{x_2' - x_1'}{K} = \frac{ct/2}{\left(1 - \frac{v^2}{c^2}\right)^{-\frac{1}{2}}}$$

$$= \frac{ct}{2}\left(1 - \frac{v^2}{c^2}\right)^{\frac{1}{2}}$$

$$= \frac{ct}{2}\left(1 - \frac{v^2}{4c^2}\right)^{\frac{1}{2}} \qquad ...[\because v = \frac{c}{2}$$

that is, in one sec he will be $\left(\frac{c}{2}\right)\left(1 - \frac{c^2}{4c^2}\right)^{\frac{1}{2}}$ or only 0.433 c away from the stationary observer's point of view.

Example: *Two particles in a given frame have velocities u_{x1} = —0.9 c and u_{X2}= +0.9 c. What is the velocity of one particle relative to the other ?*

Let us choose a reference frame moving with, say, the first particle. If this frame is called the *F* frame, then relative to this in the notation of equation 18 (*a*) the given frame has a velocity $v = -(-0.9c) = +0.9c$ and $u_{x_2} = w = +0.9e$.

Applying equation 18 (*a*) we obtain the velocity of the second particle relative to the first being equal to

$$u = \frac{v + w}{1 + \frac{vw}{c^2}} = \frac{+0.9c + 0.9c}{1 + (0.9c)0.9c)/c^2}$$

$$= \frac{+1.8c}{1 + 0.81} = \frac{1.80}{1.81} \text{ c} = 0.9994 \text{ c}$$

According to classical mechanics the relative velocity

$$u = v + w = +0.9c + 0.9c = 1.8c.$$

This exceeds *c*.

The *relativistic* velocity, however, is always less than *c*.

Theory of Waves

This theory was proposed by a Dutch citizen named Huygens, a contemporary of Newton, about the year 1678. According to him light is a form of wave-motion which is produced by mechanical vibrations of an all-pervading homogeneous medium called *luminiferous ether* just like those in solids and liquids. This artificial medium is supposed to be endowed with the properties of *elasticity* and *inertia* being thus able to transmit a wave motion. When these waves enter our eyes they excite the sensation of vision.

According to Huygens these waves were longitudinal and, as we would see later, he could quite successfully explain the phenomena of reflection, refraction, interference and diffraction Dispersion could also be interpreted if the velocity in a medium depended on the wavelength but Huygens could not account for the *approximate* rectilinear propagation of light and the phenomenon of polarisation.

Fresnel and Young overcame this difficulty by assuming that light waves were transverse in nature. Fresnel explained the approximate rectilinear propagation of light by making use of secondary wavelets and dividing the wavefront into half-period zones, known as Fresnel's zones. To explain the transmission of transverse waves ether was supposed to possess the properties of an elastic solid. Since the velocity of a wave through an elastic medium should be equal to the square root of the elasticity divided by the density and for light this should equal 3×10^8 m/sec., it follows that ether must possess a very high elasticity and a very low density. It must be able to transmit the tiniest light impulse through vast interstellar spaces, yet it must in no way be disturbed by the earth whirling through it at a speed of about 30 kilometres per second. Obviously enough, fluids like water and air do not possess these properties. Thus there is no way out except assuming that ether is a solid. The velocities of transverse and longitudinal waves in a solid medium are given by

$$v_T = \sqrt{\frac{n}{d}}$$

and
$$v_L = \sqrt{\frac{\left(k + \frac{4}{e} n\right)}{d}},$$

according as the strain is a shear or a uniform compression. Here n is the rigidity modulus, k the bulk modulus and d the density of the medium. There has been found no evidence of the existence of longitudinal disturbance in ether. To get over these difficulties ether is supposed to be highly "incompressible", *i.e.* $k = \infty$, so that, longitudinal vibration travelled with infinite velocity. On the other hand, Lord Kelvin suggested that $vL=0$, therefore, $k + \frac{4}{2} n = 0$, or $k = -\frac{4}{3} n$ *i.e.*, k is negative so that ether is "contractile". Thus, it is difficult to conceive a medium which can be stable under these conditions of contradictory properties which should be different in different media. However, the concept of elastic solid ether was very helpful in the theoretical development of the subject of light during the nineteenth century and it was with the advent of Maxwell's famous electromagnetic theory (1873) that the elastic solid theory was discarded, Till later in the last decade of the century certain other phenomena were discovered which demanded a quantum theory as proposed by Max. Planck, according to which light consists of discrete bundles of energy called *quanta* or *photons.*

For the purpose of our present discussion, we shall accept the wave-theory of light and begin with explanation of wavefronts and Huygen's principle of wave propagation.

Wavefronts and Rays: It is very often convenient to represent a train of waves by means of wavefronts. Imagine a point source of light sends out waves in luminiferous' ether supposed to be homogeneous, isotropic and all -pervading medium in which matter exists. These waves will travel in all directions with the same velocity *c, i.e.,* the velocity of light and will reach a the particles around the source lying on the surface of a sphere drawn with the point as centre simulating easily. The surface of such a sphere is known as the wavefront and all the ether particles forming the wavefront to are in the same phase of vibration. Thus the waveforms, at any instant of time, is defined as the locus of all the particles in the me sum which are being disturbed at the same instant of time and are in the same phase of vibration. The intense cottons of a series of wavefronts emanating from a point source in the plane of the diagram. As the wave advances, the wavefronts move parallel to themselves in a direction perpendicular to them. At a great distance from the source the radii of curvature of the spherical wavefronts become large and small parts of the wavefronts can be treated as almost plane and we have a train of planewaves. The normal to the wavefront are known as rays. In the case of a train of plane waves the rays are all parallel to one another.

Wave Theory: Lens Formula

Convex Lens: Suppose O is a luminous point on the principal axis of a convex lens, and XAY, the section of an incident spherical wavefront originating from O and just touching the lens surface at A. The secondary wavelets starting from the wave surface AT travel faster through air and slower through the denser medium like glass, so that during the time the secondary wavelets from A reach B through the lens, those from X and Y reach X_1 and Y_1, respectively. Thus $X_1 BY_1$ form the trace of the refracted wavefront which is assumed to be spherical and converges to the point I which forms the real image of the point O. Since the lens is *thin*, the paths XEX_1 and YFY_1 described in air are nearly straight an 1 parallel to the principal axis.

If V_1 and V_2 be the velocities of light in air and the material of the lens, respectively and μ the refractive index of the lens material, then

$$\frac{V_1}{V_2} = \mu.$$

Since the time taken by the wavelets to travel through the material of the lens from A to B is the same as that taken by light from X to X_1 or Y to Y_1 in air, therefore, this time is given by

$$\frac{AB}{V^2} = \frac{XX_1}{V_1} = \frac{YY_1}{V_1}.$$

Draw XL, EM and X_1N perpendiculars to the principal axis of the lens, then $XX_1 = YY_1 = LN$.

$$\therefore \qquad \frac{AB}{V^2} = \frac{LN}{V^1}$$

$$\text{or} \qquad LN = \frac{V_1}{V_2} AB = \mu\ AM$$

$$\text{or } LA + AB + 3N = \mu\ AB$$

$$\text{or} \qquad LA + BN = (\mu - 1)\ AB$$

$$\text{or} \qquad LA + BN = (\mu - 1)\ (AM + MB) \qquad ...(i)$$

If R_1 be the radius of curvature of the first surface EAF at which the light enters the lens and R_2 that of the second surface EBF at which it emerges and AO and BI are represented by *u* and *v*, respectively, then denoting *XL* (=*EM*=X_1N) by *y* and applying Sagitta theorem, we have

$$AL = \frac{y^2}{2u}, \; AM = \frac{y^2}{2R_1},$$

$$BM = \frac{y^2}{2R_2} \text{ and } BN = \frac{y^2}{2v}$$

Substituting the values in equation (*i*) above, we have

$$\frac{y^2}{2u} + \frac{y^2}{2v} = (\mu - 1)\left(\frac{y^2}{2R_1} + \frac{y^2}{2R_2}\right)$$

or $$\frac{1}{v} + \frac{1}{u} = (\mu - 1)\left(\frac{1}{R_1} + \frac{1}{R_2}\right).$$

According to the sign conventions used in the text, *v* and R_1 are *positive* and *u* and R_2 are *negative*, so that

$$\frac{1}{v} - \frac{1}{u} = (\mu - 1)\left(\frac{1}{R_1} + \frac{1}{R_2}\right).$$

Concave Lens: Let a spherical wave *XAY*, diverging from a luminous point *O* on the principal axis of a thin concave lens touch the lens surface at *X* and *Y*. The concave lens being thinner at the centre than at the edges and also the velocity of light in the medium being less than that in air, during the time the disturbance travels from *X* to X_1 or *Y* to F_1 in the medium the wavelets starting from the point *A* will reach the point *B* travelling partly through air and partly through the medium. Thus $X_l BY_l$ is the trace of the refracted wavefront which is assumed to be spherical and appears to diverge from the point *I* which gives the position of the virtual image of the object point *O*.

Draw *XL* and $X_1 M$ perpendiculars to the principal axis of the lens. Let V_1 and V_2 be the velocities of the light in air and the medium, respectively.

$$\therefore \quad \frac{XX}{V^2} = \frac{AC}{V_1} + \frac{CD}{V_2} + \frac{DB}{V_1}$$

Since $$XX_1 = LM = LC + CD + DM$$

$$\therefore \quad \frac{LC + CD + DM}{V_2} = \frac{AC}{V_1} + \frac{CD}{V_2} + \frac{DB}{V_1}$$

or $$\frac{LC}{V_2} + \frac{CD}{V_2} + \frac{DM}{V_2} = \frac{AC}{V_1} + \frac{CD}{V_2} + \frac{DB}{V_1}$$

or $$\frac{LC}{V_2} + \frac{DM}{V_2} = \frac{AC}{V_1} + \frac{DB}{V_1}$$

or $$= AC+DB = \frac{V_1}{V_2}(LC + DM),$$

or $$LC - LA + DM + MB = \mu (LC + DM),$$

or $$MB - LA = \mu (LC + DM) - (LC + DM)$$

or $$MB - LA = (\mu - 1)(LC + DM) \quad \ldots (i)$$

If R_1 be the radius of curvature of the surface XCY at which the light enters the lens and R_2 that of the surface $X_1 DY_1$ at which it comes out, and AO and BI are denoted by u and v, respectively, then denoting XL $(=X_1 M)$ by y, we have by sagitta theorem,

$$LC = \frac{y^2}{2R_1}, \quad DM = \frac{y^2}{2R_2},$$

$$LA = \frac{y^2}{2u} \text{ and } BM = \frac{y^2}{2v}.$$

Substituting the values in equation (i) above, we get

$$\frac{y^2}{2v} - \frac{y^2}{2u} = (\mu - 1)\left(\frac{y^2}{2R_1} + \frac{y^2}{2R_2}\right)$$

$$\frac{1}{v} - \frac{1}{u} = (\mu - 1)\left(\frac{1}{R_1} + \frac{1}{R_2}\right).$$

According to the sign conventions u, v and R_1 are negative while R_3 is positive, so that

$$\frac{1}{-v}-\frac{1}{-u} = (\mu - 1)\left(-\frac{1}{R_1}+\frac{1}{R_2}\right)$$

$$\frac{1}{v}-\frac{1}{u} = (\mu - 1)\left(\frac{1}{R_1}-\frac{1}{R_2}\right)$$

which is the same relation as deduced in the previous case.

Thus the formula

$$\frac{1}{v}-\frac{1}{u} = (\mu - 1)\left(\frac{1}{R_1}-\frac{1}{R_2}\right)$$

holds good both for a convex and a concave lens.

If the object lies at infinity, its image is formed at the focus of the lens, that is

when $$\frac{1}{u} = \frac{1}{\infty} = 0.\ \frac{1}{v} = \frac{1}{f}$$

Hence, $$\frac{1}{f} = (\mu - 1)\left(\frac{1}{R_1}-\frac{1}{R_2}\right).$$

Example: *A convex lens of focal length 20 cm and made of glass of refractive index 1.5 is immersed in a liquid of refractive index 1.33. Find the change in the focal length of the lens.*

Here refractive index of glass, ${}^a\mu_g = 1.50$

Refractive index of liquid, ${}^a\mu_i = 1.33.$

∴ refractive index of glass with respect to the liquid,

$${}^l\mu_g = {}^l\mu_i \times {}^a\mu_g = \frac{{}^a\mu_g}{{}^a\mu_1} = \frac{1.50}{1.33} = 1.128.$$

Focal length of the convex lens in air, $f = 20$ cm.

If F denotes the focal length of the lens in liquid and R_1 and R_2 the radii of curvature of its surfaces, then

$$\frac{1}{f} = (1.50 - 1)\left(\frac{1}{R_1} - \frac{1}{R_2}\right)$$

and $$\frac{1}{F} = (1.128 - 1)\left(\frac{1}{R_1} - \frac{1}{R_2}\right)$$

$$\therefore \quad \frac{F}{f} = \frac{1.50 - 1}{1.128 - 1} = \frac{0.50}{0.128},$$

or $$F = \frac{0.50}{0.128} \times 20 \text{ cm}$$

$$= 78.12 \text{ cm.} \qquad \ldots [\because f = 20 \text{ cm}]$$

Thus the change in focal length=(78.12 – 20.0)=58 – 12 cm.

Law of Huygen

Huygens' principle of wave propagation is a geometrical method of finding form the known shape and position of a wavefront at some instant, the shape and position it will have at some instant later. It is based on two postulates:

(i) Each point on the wavefront is the centre of a new disturbance called secondary wavelet. These wavelets travel out with the same velocity as the original wave os long as the medium remains the same.

(ii) The envelope, viz., the tangential plane to these secondary wavelets constitutes the new wave front.

Let us now explain the propagation of a wave on the basis of these postulates.

AB represents the trace of a spherical wavefront in the plane of the paper at any instant due to a point source of light S in a homogeneous and isotropic medium. It is required to find its position an instant t later. According to Huygens each point on the wavefront AB is the centre of a secondary wavelet.

These wavelets advance with the same velocity in the homogeneous and isotropic ether as the original. In time *t*, after the disturbance has reached *AB*, each of these secondary wavelets will have moved a distance *ct*, where *c* is the velocity of propagation. To obtain the position of new wavefront describe a series of circles with centres *a*, *b*, *c*, etc., and radii *et*. The arc A_1B_1 in the plane of the paper enveloping these circles gives the trace of the new wavefront- All the points on the wavefront A_1B_1 *at any* instant, will be in the *same* phase of *vibration*. If the source of light is at a great distance, it will give rise to a plane wavefront *AB*. A new wavefront A_1B_1 can be drawn in the same manner as just explained.

It may be noted here that in accordance with Huygens' first postulate not only we also the *backward wavefront A'B'*. But Huygens further postulated *that the action of the secondary wavelets was confined to the points at ' which they touched the forward part of their envelope*. He thus considered only the forward wavefront and assumed that no backward wavefront, as such existed.

There was, however, no direct physical or mathematical justification for this arbitrary assumption to neglect all the unwanted parts of the secondary wavelets which tend to indicate that a plane wave front spreads out slightly at the edges giving rise to diffraction effects. The absence of a back-wave was explained many years later on Stokes' law according to which the amplitude propagated in any direction due to the secondary wave is proportional to $(1+\cos\theta)$, where θ is the angle between the wave normal and the direction of propagation. For the 'backward' wavefront $\theta = \pi$ and hence

$$(1 + \cos\theta) = (1+\cos\pi) = (1-1) = 0$$

and so its effect is zero.

The waves of different wavelengths constituting a light beam, travel with the same velocity and so all the constituent waves coincide and move on as a single wave until the beam comes across a refracting medium through which the constituents travel with different velocities and undergo dispersion. This produces separate wavefronts which also travel with different velocities. Accepting Huygens' construction of the wave front we can satisfactorily explain the phenomena of reflection and refraction of light on the wave theory. But the explanation of rectilinear propagation of light on Huygens' principle of wave envelopes is not adequate, so a modified form of Huygens' principle as suggested by Fresnel is adopted.

Reflection of a Plane Wave at a Plane Surface: Suppose ADB, is the trace of a plane wavefront travelling towards a plane reflecting surface XY in the direction AA_1 and meeting it obliquely at A, E and B_1 then a, b, c normals to the wavefront are the rays of light. The incident wavefront AB and the reflecting surface XY are both perpendicular to the plane of the paper.

According to Huygens' Principle, when the wave reaches the surface, the secondary wavelets are immediately formed at the points touched so that every point of the surface XY in turn becomes the centre of secondary disturbance which is sent back into the medium. The wavefront first strikes the surface at A and lastly at the point B_1.

During the time the point B of the wavefront AB travels from B to B_1 (say t seconds) the secondary wavelet originating from A will have moved in the first medium a distance $Vt=AG = AA_1$ where V is the velocity of light in the first medium.

From B_1 draw a tangent plane to this wavelet. This will be the reflected wavefront if all the wavelets originating between A and B_1 touch this plane.

Now, in the absence of the surface XY, the wavefront AB would have occupied the position A_1B_1 after time t taken by the light to travel a distance equal to BB_1. The point D_x in A_1B_1 corresponds to the point D in AB. Thus ED_1 is the radius of the wavelet starting from E. To prove it draw EF perpendicular to B_1G, then in right-angled Δ 8 AGB_1 and $AA_1\, B_1$ side $AA_1 = AG$, AB_1 is common to both. Hence, Δ s are congruent and $\angle AB_1\, A_1 = \angle AB_1\, G$. Also the right-angled Δs EFB_1 and ED_1B_1 are congruent, for EB_1 is common to both and $\angle D_1B_1E = \angle FB_1E$ just proved.

Hence, $EF = ED_1$, thus the disturbance from E touches the tangent plane B_1G and so is the case with all wavelets originating between A and B_1. Hence, B_1G is the trace in the plane of the paper of the reflected wavefront.

Thus, a *plane wave is reflected as a plane wave from a plane surface.*

Since in right-angled Δs $AB_1\, A_1\, AB_1G$ and AB_1B,

$$AA_1 = AG = BB_1 = Vt,$$

and AB_1 is common to all.

$\therefore$ the three triangles are equal in all respects.

Hence, $\angle BAB_1 = \angle GB_1A$

Since $\angle BAB_1$ is the angle between the incident wavefront and reflecting surface so it is equal to the angle between the incident ray and the normal, *i.e.*, it is equal to the angle of incidence (i). Similarly $\angle GB_1A$ is equal to the angle of reflection (r), that is

$$\angle i = \angle r.$$

Thus, the angle of incidence is equal to the angle of reflection.

Since AB, AB_1 and B_1G are all traces in the plane of paper of planes at right angles to the plane of the paper, the normal to them all lie in the *same* plane. As the normals to $A\ B$, B_1G and AB_1 are the incident ray, the reflected ray and the normal to the reflecting surfac, respectively, so it can be stated that the incident ray, the reflected ray and the normal to surface at the point of incidence all lie in the same plane. Thus the wave theory accounts for the laws of reflection.

Reflection of a Spherical Wave at a Plane Surface: Suppose XY, is the section of a plane reflecting surface at right angles to the plane of the paper, and ACB is the section of the incident spherical wave issuing from a point source of light O and touching the plane XY at C. By the time the waves from A reach A_1 and those from B reach B_1, the waves from C would have reached C_1 if the reflecting surface were not there. Now because of the presence of the reflecting surface the secondary wavelets originate from C and will have a radius equal to CD ($=CC_1$) and reach D. In turn each point in the wavefront ACB reaches the surface and sets up a disturbance. It is evident that these wavelets will touch the surface A_1DB_1 which becomes the position of the reflected wavefront with its centre of curvature at I which represents the *virtual* image of the point source O.

Since the wavefronts $A_1C_1B_1$ and A_1DB_1 are exactly similar in curvature to the wavefront ACB,

$$\therefore \qquad OC_1 = ID,$$

$$\text{or} \qquad OC + CC_1 = IC + CD,$$

$$\text{i.e.,} \qquad OC = IC$$

$$(\text{since } CC_1 = CD).$$

Hence, the *reflected image is as far behind the reflecting sur face as the object is in front of it.*

Sagitta Theorem: If AMB, be a small arc of a circle of radius r and N is the mid-point of the chord AB, then MN is called the sagitta of the arc AMB. By geometry

$$AN^2 = MN \times NL$$

$$= MN\,(ML - MN)$$

$$= MN \times ML \text{ approx.,}$$

when MN is small.

$$\text{So that} \qquad MN = \frac{AN^2}{ML} = \frac{AN^2}{2r}.$$

As the curvature of an arc is reciprocal of the radius it follows that for spherical arcs on a common chord *AB*, for example, the sagitta *MH* is directly proportional to the curvature of the corresponding arc *AMB*.

Reflection of a Spherical Wave at a Spherical Surface: Suppose *XPY*, represents the section of the concave spherical reflecting surface with pole at *P* and centre of curvature at C. Let *XAY* be the section of an incident spherical wavefront issuing out from a point source of light *O* and touching the spherical surface at *X* and *Y*. The points *X* and *Y* become the sources of secondary wavelets which are sent back into the medium. By the time *A* wave at a spherical surface reaches *P* (say it takes *t* seconds) the secondary wavelets from *X* and *Y* reach *D* and *E*, respectively so that $XD = YE = PA = Vt$. Similarly, the secondary wavelets from all other points in turn reach the corresponding points on the surface *DPE*, so that *DPE* is the section of the reflected wavefront with its centre of curvature at *I* which also gives the position of the *real* image of the point source *O*. Draw *DL* and *XM* perpendiculars to the principal axis *OP*.

Let *PO*, *PI* and *PC* be denoted by *u*, *v* and *R*, respectively and *XM* and *DL* by *y*. Also since *XPY*, *XAY* and *DPE* are the arcs of circles of radii *R*, *u* and *v*, respectively, therefore, by Sagitta theorem

$$PM = \frac{XM^2}{2PC} = \frac{y^2}{2R},$$

$$AM = \frac{XM}{2PO} = \frac{y^2}{2u}$$

and $$PL = \frac{DL^2}{2IP} = \frac{y^2}{2v}. \qquad [\therefore DL = XM]$$

Now, $$PL = PM + ML$$

$$= PM + XD = PM + PA$$

$$= PM + PM - AM = 2PM - AM$$

$$\therefore \quad PL + AM = 2PM.$$

Substituting the values of *PL*, *AM* and *PM* from above, we get conventionally.

$$-\frac{y^2}{2v}-\frac{y^2}{2u} = -\frac{2y^2}{2R}$$

or
$$\frac{1}{v}+\frac{1}{u} = \frac{2}{R},$$

which is a relation between the distances of the object and image from the reflecting surface and its radius of curvature.

Refraction of a Plane Wave at a Plane Surface: Suppose *AB*, represents the section of a plane wavefront incident obliquely at an angle i on the plane surface of separation *XY* of the two media of refractive indices μ_1 and μ_2. Let V_1 and V_2 be the velocities of light in the two media, respectively.

The wavefront *ADB* first of all touches the refracting surface at the point *A* and lastly at the point B_1. *AB* and *XY* represent only the traces of the planes at right angles to the plane of the paper.

According to Huygens' principle each point of the surface from *A* to B_1 in turn becomes the source of secondary wavelets as the wavefront *AB* sweeps over the surface *XY*. It is evident that in the absence of any change in the medium, the disturbance from *A* would have reached A_1 during the time the disturbance from *B* reached B_1 so that $AA_1=BB_1$ and A_1B_1 would have represented the trace of the wave front after a time t which the point *B* takes to reach B_1.

Now as soon as the wavefront reaches *A* the secondary wavelet starting from *A*, has travelled into the medium for t seconds when it is just going to start from B_1. To find the position of the refracted wavefront, with centre *A* and radius $V_2 t$ describe a sphere. Draw through B_1 a tangent plane to this sphere at right angles to the plane of the paper.

If B_1G is the trace of this plane in the plane of the diagram, we have to show that B_1G is the refracted wavefront, that is

EF the perpendicular from E on B_1G is the radius of the wavelet from E when the surface is affected by the wavefront ADB.

Obviously, $AA_1 = V_1t$ and $AG = V_2t$.

$$\frac{AA_1}{AG} = \frac{V_1 t}{V_2 t} = \frac{V_1}{V_2} \qquad ...(i)$$

Again from similar Δs B_1AA_1 and $B_1 EE_1$ and Δ s $B_1 AG$ and $B_1 EF_1$

$$\frac{EE_1}{AA_1} = \frac{B_1 E}{B_1A} = \frac{EF}{AG}$$

$$\text{Consequently} \frac{EE_1}{EF} = \frac{AA_1}{AG} = \frac{V_1}{V_2}. \qquad ... (ii)$$

$$\text{But } \frac{\textit{radius in medium } \mu_1}{\textit{radius in medium } \mu_2} = \frac{V_1}{V_2}.$$

and EE_1 is obviously the radius of the wavelet in the absence of the surface XY, *i e.*, in the medium μ_1 Thus EF is the radius of the wavelet in the medium μ_2 and as it is drawn perpendicular to B_1G the wavelet must touch this plane. Similarly, we can show that all the wavelets originating in AB_1 will touch the plane B_1G, which is therefore, the refracted wavefront.

It should follow from similar reasoning as in reflection that the incident ray and the refracted ray lie in the same plane with the normal to the surface.

Moreover, $\angle$s BAB_1 and $GB_1 A$ are the angles of incidence and refraction, respectively since they are complements to the angles which the rays make with the surface, then

$$\frac{\sin i}{\sin r} = \frac{BB_1}{AB_1} \Big/ \frac{AG}{AB_1}$$

$$= \frac{BB_1}{AG} = \frac{V_1}{V_2} = \text{constant}.$$

We thus find that the sines of the angles of incidence and refraction bear a constant ratio to each other which proves the Snell's law of refraction. This ratio is termed the refractive index of the second medium with respect to the first and is denoted by ${}^1\mu_2$.

$$\therefore \quad {}^1\mu_2 = \frac{V_1}{V_2}$$

$$= \frac{\textit{Velocity of light in the first medium}}{\textit{Velocity of light in the second medium}}.$$

Further, when $r<i$, that is, the ray bends towards the normal during refraction, the ratio $\frac{\sin i}{\sin r}$ is greater than unity and the second medium is said to be more highly refracting than the first.

From the above equation it follows that the velocity of light in a highly refracting medium such as water is less than that in air or in vacuum, which is borne out by the experiments of Foucault and Michelson.

Refraction of a Spherical Wave on a Plane Surface: Let *APB*, represent the section of an incident spherical wave diverging from a luminous point *O*. When the wave reaches the boundary of separation *XPY* of the two media having refractive indices μ_1 and μ_2, respectively, then according to Huygens' construction secondary waves start in turn from all points as the wave passes over the surface. Had the wave continued in the same medium, then at some subsequent time *t* it would have occupied the position *XCY*, such that

$$PC = AX = BY.$$

But on account of the change in the medium as the point *A* moves to *X* and *B* to *Y* the point *P* moves to *D* only because the velocity of the wave inside the medium has changed from V_1 to V_2 Vi and V_2, respectively being the velocities of light in the two media. Thus the curvature of the wave front inside the medium has changed from XCY to XDY so that wave at a plane surface, the waves in the second medium appear to diverge from the point I which forms the image of the luminous point O.

Let PO and PI be represented by u and v, respectively. Drop AL perpendicular on the principal axis and represent it by y.

Since APB and XDY form the arcs of circles of radii OP and ID, respectively.

$$AL^2 = (2OP - PL)\ PL$$

$$= 2OP \times \text{PL...approx.}$$

or $$XP^2 = 2OP \times PC,$$

.... [since $AL = XP$, and $PL = AX = PC$

also $$XP^2 = (2ID - PD)\ PD,$$

or $$XP^2 = 2IP \times PD,$$

...[PD is very small, so that $ID = IP$]

$\therefore$ $$2OP \times PC = 2IP \times PD,$$

or $$u \times PC = v \times PD$$

or $$\frac{v}{u} = \frac{PC}{PD} - \frac{V_1 t}{V_2 t} = \left(\frac{\mu_2}{\mu_1}\right)$$

If $$\mu_1 = 1 \text{ and } \mu_2 = \mu, \text{ then}$$

$$\frac{v}{u} = \mu,$$

which is an important relation in geometrical optics.

Refraction of a Spherical Wave at a Spherical Surface: Suppose XAY, is the section of a spherical wave diverging from a luminous point O on the axis of a concave spherical surface XPY which separates the two media of refractive indices μ_1 (air) and μ_2 (glass). Let the wave touch the surface at X and Y.

If V_1 and V_2, respectively be the velocities of light in the two media, then

$$\frac{V_1}{V_2} = \frac{\mu_2}{\mu_1} = \mu. \qquad [\text{if } \mu_2 = \mu, \text{ and } \mu_1 = 1]$$

According to Huygens' construction when the wavefront touches the surface of separation, each point on the surface, in turn, becomes the source of secondary disturbance. The secondary wave lets from the points X and Y travel into the medium white the waves from other points on the wavefront first travel in air and then in the medium. By the time the disturbance from the point A teaches the pole P of the concave surface (where it enters the medium), the wavelets from X and Y reach B and D, respectively into the medium so that BPD gives the section of the refracted wave front which appears to diverge from the point I which is the image of the luminous point O.

Since the time taken by the disturbance in going from A to P in air is the same as that taken by it from X to B or Y to D in the medium, so that

$$\frac{PA}{V_1} = \frac{XB}{V_2}$$

$$\therefore \qquad BX = \frac{V_2}{V_1}\,PA = \frac{\mu_1}{\mu_2}\,PA \qquad \text{...(i)}$$

Draw XL and BM, perpendiculars to the principal axis OP and represent XL and BM by y, PO by u, PI by v and PC by R.

Since XAY, XPY and BPD form arcs of circles with radii OA, CP and IP, respectively, therefore, by Sagitta theorem

$$AL = \frac{XL^2}{2AO} = \frac{XL^2}{2PO} = \frac{y^2}{2u},$$

$$\text{...}[\because AO = PO \text{ approx.}$$

$$PL = \frac{XL^2}{2PC} = \frac{y^2}{2R}$$

and $$PM = \frac{BM^2}{2PI} = \frac{y^2}{2v}$$

Now, $$PM = PL - ML$$

$$= PL - XB \qquad \text{... } [\because ML = XB]$$

$$= PL - \frac{\mu_1}{\mu_2} PA$$

...[by equation (*i*) above

or $$\mu_2 PM = \mu_2 PL - \mu_1 PA$$

$$= \mu_2 PL - \mu_2 (PL - AL)$$

$$\text{... } [\because PA = PL - AL]$$

$$= (\mu_2 - \mu_1) PL + \mu_2 AL$$

or $$\mu_2 PM - \mu_1 AL = (\mu_2 - \mu_1) PL$$

If $\mu_2 = 1$ and $\mu_2 = \mu$, then

$$\mu_2 PM - AL = (\mu - 1) PL$$

or $$\mu \frac{y^2}{2v} - \frac{y^2}{2u} = (\mu - 1) \frac{y^2}{2R}$$

or $$\frac{\mu}{v} - \frac{1}{u} = \frac{\mu - 1}{R}.$$

According to the sign conventions R, v and u are all *negative.*

$$\therefore \qquad \frac{u}{-v} = \frac{1}{-u} = \frac{\mu - 1}{-R}$$

or $$\frac{\mu}{v} - \frac{1}{u} = \frac{\mu - 1}{R}.$$

If we proceed on the same lines, a similar relation is obtained for a convex spherical surface which is left as an exercise for students.

Total Internal Reflection on Wave Theory: The phenomena of total internal reflection can be satisfactorily explained on the wave theory. Let AB, be a plane wavefront in a denser medium (glass) incident at A separating it from air. Then as the wave-form sweeps over the interface the points between A and C are successively affected and become the origins of secondary wavelets in the upper and the lower media. The radius of the wavelet from A in the upper medium at the instant when B reaches C, is given by

$$AG = \frac{BC}{V^2} \times V_1 = \frac{V_1}{V_2} \; BC = \mu \; BC,$$

where V_1 and V_2 are the velocities of light in air and glass, respectively and μ is the refractive index of glass.

But $$BC = AC \sin i,$$

where i is the angle of incidence,

$$\therefore \qquad AG = \mu \; AC \sin i. \qquad ...(i)$$

Now if we vary the angle of incidence, the following three cases arise:

(*i*) When $\mu \sin i < 1$, $AG < AC$ by equation (*i*) so that a tangent plane CG can be drawn from C to the secondary wavelets originating between A and C. CG, therefore gives the refracted wavefront.

(*ii*) When $\mu \sin i = 1$, $AG = AC$ by equation (*i*). This is the *limiting case* when a wavefront is formed. The secondary wavelet originating from A will pass through C at the instant when the latter is just affected and so is the case with all secondary wavelets originating between A and C. The refracted wavefront CD, therefore, is a plane perpendicular to the surface AC proceeding in a direction parallel to the surface of separation, the refracted ray grazing the surface. The angle i denned by $\sin i = 1/\mu$, is the critical angle for the more refracting medium.

(*iii*) When $\mu \sin i > 1$, that is, the angle of incidence is greater than the critical angle, $AG > AC$. No tangent plane can be drawn from the point C to any of the wavelets originating between A and C as the point C lies within them. Therefore, a refracted wavefront cannot exist and we get total internal reflection, the reflected wavefront CD proceeding in the direction of the arrow in the same medium.

Energy Propagation

From the experiments that light travels with a finite velocity (3×10^8 m/sec.), produces heating effect, affects the photographic plates and influences the growth of plants, etc., it becomes evident that light, like heat, is a form of energy. The association of light with energy is also confirmed by the fact that it exerts pressure upon the surface on which it falls. Maxwell predicted it on theoretical grounds and Lebedew verified it experimentally in 1901. It has been clearly established by the experiments of Nichols and Hull that the energy contained in unit volume of the incident beam is equal to the pressure of light in dynes per cm^2. That light exerts pressure is also borne out by the fact that the tails of comets are always directed away from the sun. It is because the sun's rays exert pressure. Since light is a form of energy, the problem of its transmission from one place to another is purely a problem of transmission of energy. There seem to be only two modes of propagation of energy.

1. The simplest one is to suppose that the energy is carried by streams of small material particles travelling with finite velocity, and possessing kinetic energy but *not* requiring any material medium for their propagation. This is the basic principle of Newton's emission or corpuscular theory of light.

2. The other method suggests the transference of light energy by wave motion without the actual travelling

of the matter. This process requires presence of an intervening medium as a vehicle for the transmission of energy. This gave rise to the wave theory of light which was put forward by Huygens in 1678.

The Emission or Corpuscular Theory: This theory was due to Newton, the famous English philosopher (1642-1727). According to him a luminous body emitted very *minute* and weightless particles, called corpuscles, which travel through empty space (a homogeneous medium) in straight lines in all directions with the speed of light (3×10^8 m/sec.) and carry with them their kinetic energy. The particles move in straight lines simply due to property of *inertia.* When they fall upon the eye they excite the sensation of sight. Different colours of light were associated with *size* of the corpuscles. These corpuscles are considered so small that they can readily pass through the interspaces of the particles of matter. On account of their high speed and practically zero mass they are unaffected in their linear path by the force of gravity, and hence move with a uniform velocity in a straight line. This theory could easily explain some of the observed facts such as reflection, refraction and rectilinear propagation of light.

Reflection: The phenomenon of reflection of light is explained as a result of the mechanical laws of perfectly *elastic impacts,* the incident corpuscles on striking the reflecting surface bounce off at angles equal to those at which they are incident. To explain the process of reflection Newton made an *'ad hoc'* assumption that a reflecting surface exerts a force of repulsion on every panicle of light (corpuscle) incident on it in a direction normal to itself and this force decreases very rapidly with the increase of distance from the surface, so that it is effectively restricted only to a very thin layer near the surface bounded by the dotted line A_1,B_1.

Let PQ, be the path of a light corpuscle, having a uniform velocity V, incident upon the reflecting surface AB at an angle of incidence i. As the corpuscle approaches A_1B_1 that is, with in a very small distance from the surface *AB,* it begins to experience the repulsive force which continues to act, so long as it traverses this region of variable ether density, as it is generally called During its journey through this thin layer the component of the velocity of the particle parallel to the surface, *i.e., QE* (=*V* sin i) remains unaffected while the normal component *QN* (—*V cos i)* gradually diminishes, becomes zero and is finally reversed. It then gradually increases and regains its initial value when the particle reaches the boundary of this layer. Thus beyond *R* the particle moves away from the surface with the same uniform velocity *V* with which it was incident. *The path of the particle within this layer is curved.*

Since the component parallel to the surface remains unchanged during reflection, we have *QE*=*RF*

i.e., $V \sin i = V \sin r,$

or $\angle i = \angle r.$

Thus, *the angle of reflection is equal to the angle of incidence* and *by symmetry the incident ray, the reflected ray and the normal to the reflecting surface at the point of incidence are in the same plane.*

Refraction: The phenomenon of refraction of light was explained by Newton on the following assumptions:

(i) the corpuscles are attracted normally by the refracting surface *AB* of a denser medium when they come within a *very* thin layer bounded by A_1B_1 near the surface.

(*ii*) this attraction also continues to act within a *very* thin layer bounded by A_1B_2 inside the medium.

Accordingly the component of velocity normal to the surface goes on increasing during the passage of the corpuscle through the layer bounded between A_1B_1 and A_2B_2 till the corpuscle reaches the surface A_2B_2 within the medium where the attractive force ceases. After this it moves along *RS* with a uniform velocity, the parallel component of the velocity remaining unchanged. This explains the *bending* of the ray towards the normal in going from a rarer to a denser medium.

Suppose a light corpuscle is travelling along PQ. Let V_1 be its incident uniform velocity and i the angle of incidence in the rarer medium and V_2 its uniform velocity after refraction and r the angle of refraction in the denser medium. Since the components of velocities in the two media parallel to the refracting surface are the same, we have

$$QE = RF$$

i.e., $$V_1 \sin i = V_2 \sin r,$$

or $$\frac{\sin i}{\sin r} = \frac{V^2}{V_1} = \mu \text{ (a constant)},$$

which is the Snell's law of refraction.

So if $\mu > 1$ as it is for glass or water, it means that $V_2 > V_1$. Thus according to the corpuscular theory the velocity of light should be greater in an optically denser medium, like glass and water, than in a rarer one, like air.

When the second medium is rarer than the first instead of the attractive force there is the repulsive force as assumed in explaining the phenomenon of reflection. But this repulsive force is not sufficient to make the normal component of velocity zero till after the corpuscle has entered the rarer medium and reached a layer where the repulsive force ceases and thereafter the corpuscle goes in a straight line but with a diminished velocity and bends away from the normal.

But actually the magnitudes of the velocities are just the reverse as proved by the experiments of Foucault on the determination of the velocity of light in water and carbon bisulphide. His results were in full agreement with the wave theory and in complete disagreement with the corpuscular theory. This theory also failed to explain other phenomena such as *interference, diffraction* and *polarisation* of light. These shortcomings of the corpuscular theory were responsible for its downfall.

Newton's corpuscular theory, however, does not give any plausible explanation about the origin of repulsive and attractive forces normal to the surface while explaining the phenomena of reflection and refraction. In the case of transparent substances like glass, water, etc., light is partly reflected and partly refracted at the surface simultaneously. How it is possible for the surface to repel as well as to attract the particles of light at the same time ? To explain this difference in the behaviour of the corpuscles, Newton suggested the theory of fits, according to which when the corpuscles reach the surface, sometimes they are in a *fit of easy reflection,* and sometimes in *a fit of easy transmission* and pass periodically from one state to the other.

Newton also postulated the existence of a hypothetical medium, ether with variable density, being greater in free space and less in a material medium. According to him the velocity of the corpuscles depends upon the density of ether, being less in a rarer medium and greater in a denser one. Moreover, the impinging rays produce vibrations in the ether at the interfaces. These vibrations alternately contract and dilate the ether near the surface which, respectively causes the reflection and refraction of the light corpuscles.

To explain the phenomenon of dispersion, Newton assumed that corpuscles of different colours have different sizes and that in a dispersive medium velocities of corpuscles depend upon their size. So that when white light passes through such a medium corpuscles associated with different colours deviate through different extents, thus bringing about dispersion.

Inspite of several arbitrary assumptions Newton's corpuscular theory could not explain the phenomena of interference, diffraction, polarisation, double refraction and spectrometric data, etc.

Spectrum and Radiation

Invisible Radiations

We have considered above that portion of the spectrum which when incident on the eye causes sensation of sight. But extending on either side of this visible portion there are invisible radiations which are too long or too short to affect the eye but they carry energy and obey the same laws as light, can be polarised and diffracted by concave reflection gratings. The portion beyond the red consists of longer wares extending upto 400,000 Å and is called the infrared and that beyond violet consisting of shorter waves extending upto 136 Å is called the ultraviolet.

The Infrared: This part of the spectrum was first discovered in 1800 by Sir William Herschel who observed that when the blackened bulb of a thermometer was placed even at some distance beyond the red end of the spectrum, it indicated a rise in temperature. This radiation is emitted copiously by the sun and other hot bodies and contains most of the heat energy of the sun rays. It is generally detected by the *heating effect* for which we may use the blackened bulb of a thermometer or a Crooke's radiometer. But more sensitive instruments like thermopile, Langley's bolometer and Boy's radio-micrometer are now used. As a result of the study of these rays we find that the maximum heating effect is always beyond the red end of the spectrum in the case of electric lamp and electric arc and in the visible part if the sun is the source. The spectrum of this part is continuous like the visible part and contains a number of dark lines where the heating effect is nil. Most of the properties of these rays are common with the visible portion and these rays extend so far as to merge with the radio waves at about 4×10^{-2} cm.

Since 1880 the infrared portion of the spectrum has been carefully investigated and the wavelengths of lines constituting this portion have been measured. In the study of this region, we make use of a spectrometer with prisms and lenses made of *rock-salt* and *sylvine* for different ranges of wavelengths because glass absorbs them. It is also better to dispense with lenses altogether and to collimate the beam incident on the prism and to focus the emergent beam, stainless steel mirrors are employed.

The various methods that have been used for detecting and investigating infrared rays are: *i*) *Heating effects*, *ii*) *Photography with special plates*, and *iii) Photoelectric effect*. We shall now discuss these methods and the uses of these radiations.

(i) *Heating Effects — Infrared Ray Spectrometer:* To investigate the intra-red rays spectrum, we make use of the heating effects of the rays. A special type of infrared ray spectrometer is used for examining the intensity of these rays and for determining their wavelengths. A powerful source of infrared radiations illuminates the slit S_1. The radiation after reflection from the mirror C_1 is incident on a rocksalt prism P and after refraction through it the emergent beam is allowed to fall on a plane mirror M (both P and M are mounted on the rotating table). This beam after reflection from M falls on a concave mirror C and is brought focus on the detector T, which may be a *thermopile* or a *bolometer* connected to a sensitive galvanometer.

By keeping the detector fixed in position the prism table is rotated to bring the different lines in the infrared spectrum on to the detector. This is indicated by the deflection of the galvanometer connected to the thermopile which rises to maximum when the line falls on the slit S_2. Thus the spectrum of any source can be mapped out. To actually measure the wavelengths of any source, we have first *to calibrate* the instrument by taking a source of infrared radiation whose wavelengths have already been measured by a reflection grating. The readings of the prism table are taken for each known wavelength and are plotted as abscissa and the corresponding readings of the galvanometer as ordinates.

(ii) *Infrared Ray Photography Audits Uses:* Photography is another means of detecting infrared radiation and is utilised with advantage in the petrography of the rays. The serious drawback of concave reflection grating spectra is that only a small portion of the incident radiation goes into the diffracted spectra which are therefore weak. Consequently exposures extending over several hours are necessary to get the spectral lines bright enough to be measured accurately. This is done with special photographic plates.

The infrared ray photography has been very usefully employed for taking the photographs of distant landscapes in the dark even, because the rays can penetrate easily through the atmosphere. Dense fog and smoke are transparent to them. For this purpose special panchromatic photographic plates and infrared ray filters are employed. A solution of iodine in alcohol is a suitable filter. It allows to pass through infrared rays easily but is opaque to visible light.

In World War II, infrared ray photography was successfully used in revealing the positions of camouflaged buildings, in seeing through fog, in detecting enemy concentrations in the dark.

With the manufacture of special kinds of glass which is transparent only to a particular region of the spectrum the infrared rays have been successfully employed for *secret signalling*. This requires a powerful source of infrared radiation, which is so hot that it also emits light. But the light radiation which will spoil the secrecy, is cut off by the special glass screen transparent only to the infrared.

(iii) *Photoelectric Effect of Infrared Rays:* This effect relates to the emission of electrons when radiations are incident on the surface of an alkali metal and will be discussed in detail. It may be usually applied to study the infrared rays if a photoelectric cell of some suitable material such as rubidium is employed.

On account of their photoelectric effect these radiations are used in television for illuminating the objects. They are more suitable than ultraviolet radiations since the latter have harmful effect on the persons being televised.

In addition to the above applications due to their easy absorption by human body they have been used for *therapeutic purposes* such as the diagnosis of superficial tumours and varicose veins, in, curing infantile paralysis, sprains, dislocation of bones, etc.

The Ultraviolet: Beyond the violet end of the spectrum, there is the ultraviolet portion. It spreads over several octaves ending in X-rays and then in ϒ-rays. It extends right up from 4000 Å to 136 Å. This region was discovered by Hitter in 1801 while investigating the photographic action of ordinary light on silver chloride. He found that the effect did not stop at the violet end of the spectrum but extended far beyond it.

The ultraviolet radiation is reflected in the same way as visible light but some metals behave in an interesting manner towards it. A thin layer of silver, for instance, reflects 95 per cent of the incident light in the visible part of the spectrum but only 4 per cent in the ultraviolet, thus forming a very good filter for transmitting the ultra violet while intercepting the visible. Nickel, magnesium and many other alloys act in a similar way and even more effectively than silver.

In 1811, Thomas Young formed Newton's rings with ultraviolet radiation and photographed them. He found that the radii of the corresponding rings were smaller with ultraviolet light than with visible light, thus showing that their wavelength was smaller than that of visible light. Subsequent experiments with reflection gratings confirmed that ultraviolet light consisted of waves in ether differing from light only by its smaller wavelength.

The ultraviolet solar spectrum is found to be continuous but traversed by a number of dark lines like the visible spectrum. These radiations obey the usual optical laws, can be caused to interfere and can be polarised by tourmaline crystals just like visible light. Besides the sun which is the most powerful source of ultraviolet radiation, we have several artificial sources rich in ultraviolet such as are lamps with iron or carbon electrodes, mercury vapour lamps, the spark spectrum, the mercury are aclosed in a quartz tube and discharge tubes.

The chief methods for detecting and investigating ultraviolet radiations are: *i*) *Photography, ii*) *Fluorescence, iii*) *Phosphorescence,* and *iv) Photoelectric Effect.*

(i) *Ultraviolet Ray Photography:* To measure the wavelength of these rays and to investigate other properties and effects, we employ the photographic action of the rays. Since these radiations are easily absorbed by glass, the spectrographs make use of quartz and fluorspar prisms and lenses, the latter being transparent upto 1200 Å.

Concave reflection gratings have been employed for measuring their wavelengths as in the case of infrared radiation but as the diffraction spectra are very weak on account of the poor reflecting capacities of the metals for ultraviolet radiations and also the energy being spread over several spectra, prism spectrometers are generally preferred. They give comparatively much intense spectra and combined with long photographic exposures, very satisfactory results are obtained.

(a) *Ultraviolet Ray Spectrograph*: Since quartz of which the prisms and lenses are made is doubly refracting it results in doubling of each spectral line and thus decreasing its sharpness and the resolving power of the spectrometer. To eliminate this defect Cornu prism is used which is made by holding together with a thin film of glycerine two similar right angle 30° prisms each cut with its optic-axis parallel to the base. One of them is of *right-handed* and other of *left-handed* quartz so that double refractions balance and the sharp spectrum results. The collimating lens and the telescope objective are, respectively made of right-handed and left-handed quartz and are cut with the axes parallel to the optic axis of the quartz.

In order that a wide range of wavelengths may be focused on the photographic plate *G* to cover up the differences in the dispersive power it is tilted so as to be at about 20° to the axis of the lens.

(b) *Littrow Spectrograph:* A compact form of the spectrograph for precision spectrography is the one designed by Littrow and is known as Littrow spectrograph. Light from the source under examination is passed through a slit *S* and is totally reflected by a quartz prism *R*. It then fails on a lens *L* and emerges as a parallel beam to become incident on a quartz prism *Q*, whose back surface is suitably coated to make it reflecting for ultraviolet light. The prism *Q* is set in the minimum deviation position for the sodium light and the refracted beam strikes almost normally at the back surface, which reflects the light back through the prism. The dispersed beam then passes through the lens which forms an image of the spectrum on the photographic plate *P*. Different regions of the spectrum are brought to focus on the plate by rotating the prism *Q* which it mounted on a rotatable table The prism *R* is arranged in such a way that the returning rays *pass over* it to the plate *P*. To absorb reflections from the lens surface to avoid fogging of the image, a small blackened screen is placed behind *L* and *R*- The whole instrument is enclosed in a light-tight box mounted on a straight girder.

This spectrograph has the additional advantage that the optic axis of the quartz being normal to the back surface of the prism, no double refraction occurs and any splitting due to the resolution of light into two opposite circularly polarised components is exactly compensated when the ray retraces its path out of the prism again. Such exact compensation, however, is not possible with Cornu prism on account of the difficulty of preparing two prisms of identical dimensions, one of left-handed and the other of right-handed quartz. Littrow spectrograph has also been successfully employed with a plane reflection grating for the study of very short and very long wavelengths in an optical spectrum.

Various other types of spectrographs and mountings have been employed depending upon the region to be investigated as the materials used for transmission have different powers of absorption for different regions. For wavelengths lying below 1850 Å vacuum spectrographs with concave gratings have been found most suitable. With such arrangements in the years 1890-1900. Schumann investigated upto 1200 Å. This is known as the Schumann region of the spectrum. Lyman, Millikan, Seighbahn and others have photographed lines of as short wavelengths as 200 Å.

(ii) *Fluorescence:* When ultraviolet light is incident on certain substances, they emit visible light. For example, when a solution of quinine sulphate in dilute sulphuric acid is exposed to the ultraviolet portion of the spectrum it is found to glow brightly emitting blue light. This phenomenon is termed fluorescence. It is the process in which a substance absorbs a part of the incident ultraviolet light and then immediately re-emits an appreciable part of it with its wavelengths longer than those absorbed. This act is known as Stokes' law. The energy of the absorbed light raises the atoms or the molecules of the fluorescent substance from the lower to the higher energy levels and when these atoms or molecules return to their normal states light of longer wavelength is emitted and in accordance with quantum theory, the loss in energy or the energy absorbed by the fluorescent substance is $h(v_a - v_e)$, where h is the Planck's constant and e_a and v_e are the frequencies of the light absorbed and emitted, respectively. This is also what Stokes' law predicts. There are, however, cases where Stokes' law is violated because sometimes the energy of the excited atoms and molecules of the material are added to the energy of the emitted waves, the atoms and the molecules themselves returning to the normal or some lower energy level.

There is a large number of substances which exhibit fluorescence such as uranium glass, gems, eocein, kerosene oil and various lubricating oils. Paraffin oil when exposed to sunlight exhibits a blue fluorescence. An aniline derivative termed fluorescein emits a yellowish green light when exposed to day light or an electric arc.

In the commercial fluorescent lamp the inner side of the tubular bulb is coated with fluorescent material called *phosphors* which absorbs the invisible ultraviolet light from the low pressure mercury arc and emits visible light. Fluorescent screens of cathode ray oscilloscopes and television tubes and to make visible with *X*-rays the shadows of bones, are other important applications of the phenomenon of fluorescence.

It may be noted that fluorescence persists so long as the fluorescent substances remain exposed to incident ultraviolet light and re-emission of light ceases as soon as incident light is cut off.

(iii) *Phosphorescence:* There are some materials in which the molecules excited by the absorption of incident ultraviolet light do not immediately return to their original state and the emission of light continues even after the exciting radiation is removed. This type of *delayed* fluorescence is called phosphorescence. The phenomenon is caused chiefly by the ultraviolet and violet parts of the spectrum and is usually accompanied by fluorescence, which may be used as a method of detecting and investigating the ultraviolet region. The common substances which exhibit this effect permanently are sulphides of calcium, barium and strontium. Luminous paints which are composed of these sulphides combine to phosphoresce for several hours in a dark room after exposure to sunlight. The coating materials used in fluorescent tubes show this effect to some extent, as we often notice that the glow is still faintly visible for some time after the electric current is switched off.

There is, however, no *essential* difference between fluorescence and phosphorescence except that the latter persists after the exciting radiation has ceased, whereas in the former, light is re-emitted within 10^{-7} or 10^{-8} s. There are gases, liquid as well as solids such as soda glass, etc. which, when bombarded by charged particles such as electrons, emit light. They are also called fluorescent substances. The emission process is the same but the excitation of the atoms of the substance results from the electric field of the charged particles -rather than from the electric field of the light waves.

(iv) *Photoelectric Effect of Ultraviolet Rays:* The ultraviolet radiations are most effective in ejecting electrons from alkali metals which thus acquire a *positive* charge on being exposed to ultraviolet radiations so that if a plate of such a metal is joined to a quadrant electrometer or a sensitive electroscope a deflection will occur. If now a sheet of glass is interposed between the metal plate and the source of radiation, the effect may be stopped in most cases thus showing that it is due to the ultraviolet rays which are now absorbed by the glass plate.

Hydrogen Spectrum

The order in the spectral lines of hydrogen was first discovered by Balmer in the year 1885, when he was investigating its spectrum. He found, as a result of close study, that the frequencies of four lines in the visible region together with several more in the near ultraviolet could be represented by the simple formula

$$v = \quad R \ Hz, \qquad \ldots(1)$$

where $n = 3, 4, 5,$......for various lines and R is a constant having the value 3.291×10^{15} s^{-1}. It is now called Rydberg's constant, named after a Swedish spectroscopes.

If we put $n = 3$, we get H_α line, for n=4, H_β line and so on.

It is convenient to express the above formula in terms of the wavelength.

Now $\quad v = \quad ,$

$\therefore \quad = 3.291 \times 10^{15}\ s^{-1}$ ... by (1) above

or $\quad =$

... [$\because$ c, the velocity of light = 3×10^8 ms^{-1}

$= 1.09678\ m^{-1} \times 10^7$ ...(2)

The ratio $1/\lambda$ or the reciprocal of wavelength is called the wave number, that is, the number of waves per m in

vacuum and now the Rydberg constant has the value $1.09678 \times 10^7\ m^{-1}$. The wave number is denoted by v–.

Simplifying further, we have

$$v\text{–} = \quad = 1.09678 \times 10^7\ m^{-1}$$

or $\lambda =$. m

$=$ A. [$\because$ 1 m = 10^{10} Å]

$=$ 3645 Å, where n = 3, 4, 5, 6, etc.

This expression gives the wavelengths of the Balmer series in hydrogen in excellent agreement with the observed values.

We find that the lines get closer together as we go towards the shorter wavelength side crowding up to a series limit so that by putting $n = \infty$ in the formula

$v = R$,

we have for the frequency of the shortest line

$v =$ in the visible region.

This work of Balmer was extended later by other experimenters and several other spectral series have been discovered.

(*i*) *Lyman series* in the ultraviolet,

$v = R$, where n = 2, 3, 4, etc. ... (3)

The lines crowd to a series limit at $v = R$.

(*ii*) *Ritz Paschen series* in the infrared,

$v = R$, where n = 4, 5, 6, etc. ... (4)

(*iii*) *Brackett series* in the far infrared,

$v = R$, where n = 5, 6, 7, etc. ... (5)

(*iv*) *Pfund series* in the extreme infrared,

$v = R$ where n = 6, 7, 8, etc. ... (6)

More generally, all these series can be denoted by the single equation,

$v = R$ ...(7)

where n_1 and n_2 are integral numbers. For Lyman series n_1= 1. For Balmer n_1= 2 and for Paschen, Brackett and Pfund series n_1 = 3, 4 and 5, respectively.

Thus we find that the frequency of a given line can be expressed as a difference of two spectral terms represented by

R, .

These terms it appears represent the energy states or energy levels in the mechanism of radiation from an atom.

Energy Levels: A number of *energy levels* are represented by a set of horizontal lines such that their vertical distances below the line OO' are proportional to the values of the terms

$R,\quad ,\dots\dots$

Transition between any two of these energy levels gives a corresponding line in the spectrum.

Thus the frequency of the first line of Lyman series is given by the difference between R and $R/4$ levels and that of the second by the difference between R and $R/9$, that is, all the lines of the series are represented by the vertical separation between the lowest level and any of the upper levels. Similarly, for Balmer series the lines are represented by the vertical separation between the $R/4$ energy level and any one of the upper level and so on, for other series. Of course with the complex spectra the picture is not so simple as in the case of hydrogen but the analysis of the lines has shown that it is possible to represent all the lines of the spectrum as difference between different members of a set of terms. For explanation of hydrogen spectrum of Bohr's theory.

Absorption Spectra: If light from an arc lamp giving a *continuous* spectrum is examined by a spectroscope after it has passed through a gas at low pressure and lower temperature than that of the source it is found that the spectrum has certain dark regions, showing that certain wavelengths are absorbed. This is called absorption spectrum of the substance. As in the case of emission spectra, we have both the *continuous* or *general absorption* and *selective* or *line absorption.*

If red or blue glass is placed in front of the sunlight or arc light, it only allows a narrow band of red or blue rays to be transmitted giving a continuous absorption spectrum. In the case of cobalt glass, the spectrum is crossed by three dark bands indicating that it absorbs three different groups of waves. There are many substances, solids and liquids, such as potassium salts and aniline dyes which produce such spectra giving rise to isolated dark bands whose number, width and position are characteristic of the absorbing substance. It is possible, therefore, to use this property in the analysis of the solutions of such substances just like the emission spectra. A highly developed technique of absorption spectroscopy is made use of for the identification of dyes, drugs, oils, glasses, etc. On the other hand, in the case of an absorbing gas or vapour narrow sharp lines have been observed in absorption spectra as in the emission spectra, that is, absorption occurs for certain definite isolated wavelengths. This can be demonstrated by a simple experiment.

Suppose a sodium flame is interposed between the arc lamp and the slit of the spectrometer it is found that the spectrum is crossed by two dark lines in the yellow portion. These dark lines occur exactly in the same place as is normally occupied by the bright sodium lines, a fact which can be verified by holding an opaque screen in front of *S*, so that the spectrometer slit is illuminated entirely by the light from the sodium flames.

This means that sodium vapour absorbs those wavelengths which it itself is capable of emitting when acting as a luminous source. This is the absorption spectrum of the sodium lines and the phenomenon is known as the Reversal of Sodium Lines. Theoretically the phenomenon can be explained as a case of resonance as the ultimate particles absorb energy from the continuous spectrum corresponding to their own natural frequencies. Kirchhoff and Bun sen as a result of their study of absorption spectra formulated a law concerning the emission of light by substances known as Kirchhoff's law.

It states that — a substance which emits a light of certain wavelength when heated to incandescence will absorb the light of the same colour when cold if light from a source at a higher temperature is made to pass through it. It is the light corresponding to yellow lines in the arc light (higher temperature) which is absorbed by the sodium vapour in the flame (lower temperature) that we get dark lines in their place In fact these lines are not dark but comparatively less luminous than the rest of the spectrum.

Solar Spectrum and Fraunhofer Lines

When we examine solar spectrum carefully by a spectrometer, we find that it is crossed by a large number of narrow dark lines distributed throughout the whole length of the spectrum. They were first observed by Wollaston in 1802 and later studied by Fraunhofer, a German optician, in 1814, and, therefore, called after him as Fraunhofer lines.

Fraunhofer counted some 600 of these lines and measured the positions of 324 and showed that they occupied definite and constant positions in the solar spectrum. He denoted the chief lines by letters *A, B, C, D, E, F, G,* and *H,* the lines *D* corresponding to sodium, *C* and *F* due to hydrogen and so on. The interpretation of these lines remained a mystery until the work of Bunsen and Kircchoff on spectra appeared in 1859.

According to them these lines are produced by absorption in the atmosphere of the sun. The central part of the sun called the *photosphere* is intensely hot and emits a continuous spectrum. Surrounding this central part there is an atmosphere of vapour and gases called the *chromosphere* which is comparatively colder. When light coming from the interior of the sun passes through the chromosphere, its vapours absorb light of certain wavelengths which they themselves would emit, if incandescent.

Thus if the vapours of sodium, iron and copper are present in the chromosphere, white light in passing through them would lose those colours (wavelengths) which these elements would emit at the higher temperature. The corresponding wavelengths would, therefore, be missing when the solar spectrum is examined. By making a comparison of missing wavelengths with the wavelengths emitted by various elements, it is possible to discover the elements composing the sun. The relative intensity of the bright and dark lines shows the relative abundance of the elements in the chromosphere. It has been found that the sun's atmosphere consists of vapours of as many as 60 elements existing on the earth.

It may be noted that some of these lines are also due to absorption by oxygen, ozone, water vapour and carbon dioxide in the earth's atmosphere.

A table of chief Fraunhofer lines is given below in which the wavelengths are expressed in terms of Angstroms (Å).

$$1\ \text{Å} = 10^{-8}\ \text{cm.}$$

Sometimes other units are also used in the measurement of wavelengths.

A micron (1 μ) = mm = 10^{-3} cm = 10^{-6} m

A milli-micron (1mμ) = micron = 10^{-6} cm = 10^{-7} m

= 10^{-9} m

1 mμ = 10 angstroms

An Angstrom (Å) = 10^{-8} cm = 10^{-10} m

= 10^{-7} mm = 0.1 mμ.

Spectra of Heavenly Bodies: We have just discussed the solar spectrum. An examination of the spectrum of the moon reveals that it is identical with that of the sun. This is due to the fact that moon has no atmosphere of its own which could rob the sunlight reflected from it of any of the frequencies present in it. In the case of most of the other planets absorption bands are found to exist in the light they reflect. Thus carbon dioxide has been shown to occur in the atmosphere of Venus, ammonia and methane in the atmosphere of Jupiter and latter alone in that of Mars. It is also possible to estimate the amounts of these compounds from the intensities of the absorption bands in their respective *spectra.* The stellar spectra have also been photographed and they show the presence of the same elements as are found on the earth. Iron, calcium, hydrogen and titanium are prominent amongst them. The study of the stellar spectra has been useful in several other ways. We have been able to find the distances, brightness, surface temperature, radial velocity, size and mass of the stars. It is from the type of the absorbing spectrum that we can estimate the temperature of the absorbing material. For instance, by examining the relative intensities of the absorption lines of hydrogen, calcium and iron in the spectrum of *Arcturus,* the surface temperature is estimated at about 4000°C.

Table showing wavelengths of Fraunhofer lines in Å

Line	*Element*	*Wavelength*	*Line*	*Element*	*Wavelength*
A	Oxygen	7661	*h*	$H_{(\delta)}$	4102
B	Oxygen	6867	*H*	C_a	3968
C	$H_{(\alpha)}$	6563	*K*	C_a	3934
		5896			
D	*Na*	5890	..	$\mu_g - C$	3838
E	E_s	5270	*L*	$E_s - C$	3820
		5178			
b	μ_g	5173	*M*	F_s	3720
F	$H_{(\beta)}$	4861	*N*	F_s	3581
f	$H_{(\Upsilon)}$	4340	*O*	F_s	3441
G	F_s	4308	..	$T_j - F_s$	3057

The Doppler Effect in Light: We know from our everyday experience in the case of sonorous bodies that if there is relative motion between the source and the observer, an apparent change in pitch takes place. This is called Doppler effect. When the source approaches the observer, there is an apparent rise in pitch and hence a decrease in wavelength and *vice versa.* The same phenomenon is observable in light provided the relative velocity v of the stellar source and the earth along the line of sight is appreciable in comparison with the velocity of light c. The wavelength of every spectral line undergoes a change. Suppose the source is in motion. Let f be the frequency of the source, v its velocity towards the observer and c the velocity of light. Since the waves produced per second are crowded into the space (c – v) when the source approaches the observer, the wavelength of ligh will be shortened from

$$\lambda = \text{ to } \lambda' =.$$

The observed frequency of the light f' =, then

$$f' = f. \qquad \text{... (8)}$$

If the source is receding, the sign of v is reversed with the result that

$$f' = f. \qquad \text{... (9)}$$

Combining these results we have the apparent frequency

$$f' = \;\; = = = f \text{ ... (10)}$$

If the observer is in *motion* with velocity v then during approach it will actually receive more waves per second equal to v/λ and during recession the number of waves received will be reduced by v/λ, therefore, the apparent frequency,

$$f = f \pm = f \pm f = f. \qquad \text{... (11)}$$

It should be noted that in light there is no physical difference between the two cases whether the source is in motion or the observer is in motion because this distinction is based on the assumption that there is a medium, such as air, for the propagation of sound waves while no such medium has been discovered for light waves (Relativity theory).

Expanding the right-hand side of equation (10) by the binomial theorem, we have

$$f = f,$$

which is the same as equation (11), if v^2/c^2 and higher terms are neglected.

So in actual practice, equation (11) is used.

If λ' is the altered wavelength and λ the actual wavelength, the equation (11) becomes

$$= \quad \text{or} = . \qquad \text{...(12)}$$

This change in wavelength can be observed when the stellar spectra are photographed alongside that of an arc or spark giving some of the same lines as in the stellar spectrum.

If the corresponding stellar lines are shifted towards the violet end indicating a diminution in wavelength, we conclude that the star is approaching us, and if the shift is towards the red end indicating an increase in wavelength, the star is receding from us.

Hoygens detected the displacement of the *F*-line towards the red end in the spectrum of Sirius as compared with the spectrum of sun, indicating a recession of 48 km/sec. Similar observations have indicated that the stars Sirius, Rigel, Regulus and Astor are receding from the Earth while Arcturus and Vega are approaching the earth. Most of the stars have been observed to have radial velocities ranging from 10 to 300 km/sec.

There are a few other applications of the Doppler effect:

(i) *Angular Velocities of the Sun and the Planets:* By comparing the position of the Fraunhofer lines in the spectrum of two different parts of the surface of a planet at a time we may find different shifts in the two cases indicating that all the parts of the planet do not have the same velocity towards the earth. By measuring this shift angular velocities may be determined. By this method we have discovered that when the spectrum of the west edge of the sun is compared with that of the east edge the absorption lines show a displacement corresponding to a velocity difference of about 2 kilometres/sec. A similar comparison of the spectra from the north and south edges of the sun shows no such displacements. Thus we conclude that the sun rotates about an axis passing through north and south.

(ii) *Discovery of Double Stars:* Some stars which ordinarily appear single are in fact double stars known as *Spectroscope binaries.*

They revolve about each other so that when one is approaching the earth, the other is receding from it. Their spectrum on one night may show all the lines as single. This would happen when one star is covered by the other and there is no component of velocity in the line of sight.

When both the stars become visible after a few nights on account of the approach of one and the recession of the other, the displacement of the spectral lines from

.e two will be in opposite directions and double lines ·ould result. By this method we have discovered that several stars are double.

(iii) *The Red Shift:* The most important application of the Doppler effect is that the distant spiral nebula is found moving away with velocities as great as 14300 miles/ sec. Bright lines in the spectra of these island universes outside our own Galaxy (the Milky Way) appear to be displaced towards the red end of the spectrum, an effect known as the *red shift.* These observations have given rise to the belief that our universe is expanding.

(iv) *Saturn Rings:* The telescope shows three concentric rings round the planet Saturn. It has long been disputed whether these rings were solid or composed of a swarm of smaller satellites In the former case the velocity of the outer edge of a ring would be greater than that of the inner one whereas in the latter case, the satellites nearer to the Saturn would have greater velocity than those farther away to balance greater gravitational pull.

[Since for Satellites motion,

$$= \quad \therefore \quad v^2 =$$

that is, v increases as R decreases.]

This means that the innermost edges of a ring would show a larger Doppler shift than its outermost edges. This result has been confirmed by actual measurements of the shift. Therefore, it is concluded that rings are composed of swarms of individual satellites.

Ultraviolet Radiations in Operation

(1) The ultraviolet radiations within the range 2000 Å and 3000 Å are very active biologically. They kill bacteria and are very effectively used for sterilising air, water supplies and other materials in bulk.

(2) *Ultraviolet ray therapy* is now a medical technique of curing virulent ulcers, some skin affections and infected

wounds. The rays are found to be an essential element in the manufacture of vitamin *D* in the skin, so necessary to healthy life. It is for this reason that persons suffering from rickets, tubular abscesses and diseases of the bone are advised regular exposures to sunlight or artificial sources of ultraviolet rays. Miners who are deprived of the normal amount of ultraviolet radiation which come from the sun are given occasionally artificial dose of these radiations to partly make up the deficiency caused by underground living. They assist in increasing the rate at which calcium and other minerals art absorbed in the system, stimulate the nervous system and ductless glands and help oxidation of fats.

(3) The difference in the fluorescence caused by them in different types of inks enables us to detect *forged* documents. For the same reason we can distinguish between real and false gems, artificial and real teeth.

(4) Fluorescent lamps depend for their action upon the fluorescence caused by the ultraviolet light in phosphors coated inside the fluorescent tubes.

(5) We have already learnt that the resolving powers of the telescope and the microscope depend upon the wavelength of the light used, the smaller the wavelength, the greater the resolving power. So the resolving power increases if the visible light is substituted by ultraviolet light. The grain of the image improves and more details are observed Ultraviolet radiations have enabled us to photograph bacteria even.

The Complete Radiation Spectrum: In our study of the optics we have dealt with the visible, the infrared and the ultraviolet portions only. Work during the last several decades has resulted in the discovery of radiations that are hundred million times longer than the red known as radio waves, others given off in the *subatomic* transformation are ten million times shorter than the violet, all essentially resembling light in their physical nature. They all travel with the velocity of light equal to 3×10^8 ms^{-1} in vacuum.

In 1887, Heinrich Hertz succeeded in producing by electrical means waves that could be reflected, refracted, polarised, etc. like ordinary light. Their wavelength was found to be much greater than that of longest infrared rays (0.01 cm).

Subsequently radio-waves of the very highest frequency the so-called microwaves were produced with wavelengths as short as 1.8 mm and even still smaller, so that the gap between micro-radio-waves and infrared waves no longer exists.

Below the ultraviolet region we come to *X*-rays produced in a discharge tube by the sudden stoppage of the electrons emitted from the cathode when they hit a target of hard metal. X-rays too were found to possess all the properties of light waves except that they were much shorter in wavelength of the order of 0.1 Å. It is this fact which accounts for the difficulty in reflecting and diffracting *X* rays.

Similar to *X*-rays but more penetrating are the so called γ rays emitted by radioactive substances and in many subatomic processes. They are found to have wavelengths of 10^{-2} Å and less. They too have been shown to be of the same essential nature as light waves. Still shorter are the cosmic rays coming from the sun or interstellar spaces. These waves have great powers of penetration as their wavelength is extremely small of the order of 10^{-12} cm.

All these types of radiations described here have been shown to be waves of an electromagnetic character and we speak of the complete spectrum as electromagnetic spectrum. A complete chart of this spectrum and the part occupied by the visible light will be seen to be extremely small in comparison with the wavelengths known.

It may, however, be noted that while the various ranges are in general produced and detected in different ways and cause widely different effects, all consist of the waves of the same fundamental nature differing only in wavelength. The unification represented by the electromagnetic spectrum is one of the most impressive facts in physics and is the result of an enormous amount of research and investigation.

Sources of Spectra

We are all familiar with Newton's experiment which proved conclusively that sunlight consisted of an infinite number of constituents all refracted by the prism at different angles but grouped together in seven different colours, red, orange, yellow, green blue, indigo and violet.

Besides solar light used by Newton, we have now studied the spectra emitted by number of sources of light, some of which we describe below:

The Incandescent Solid

A solid body heated to a high temperature has long been used as a source of light for the study of spectra. It emits a continuous spectrum and all the colours are gradually added up as the temperature rises higher and higher. A tungsten filament lamp is nowadays a common example of such a source of light. A modern efficient variety of this lamp is the coiled coil type filled with argon and nitrogen at reduced pressure to diminish evaporation losses. The temperature in this case is as high as 2100°C.

The Bun Sen Flame

Any substance introduced in the non-luminous part of the flame on being volatilised results in the characteristic spectrum which may consist of a number of spectral lines. To introduce the salt in the flame we use an asbestos ring dipped in the solution of the salt or we take a platinum wire, heat its one end and dip it in the salt which melts and adheres to it. It is then inserted in the flame. This is our usual method of obtaining a sodium flame.

The Arc

When two adjustable carbon rods connected to electric supply mains are touched together and then slowly separated, brilliant arc is produced across the small air gap. A photographic examination of the arc reveals carbon are spectrum. When some volatile metal is inserted in the arc we will get the characteristic spectral lines of the metal. In place of carbon rods one may have metallic rods in which case the metal arc results. With metals of low melting point the arc may be maintained In an evacuated vessel. The most familiar examples of this are the mercury vapour lamp and the sodium vapour lamp.

Two forms of the mercury vapour lamp are in common use. The form (a) consists of an outer vessel *A* and an inner vessel *B*, the space between them being evacuated. The inner vessel contains a small quantity of mercury and argon; sealed at the ends are two main tungsten spiral electrodes E_1 and E_2 with a starting electrode 5 connected to the distant electrode E_2 through a very high resistance In each of the main electrodes there is a rod of barium compound which emits electrons on applying A.C. voltage to them through a choke *P* which serves as a stabiliser. To start with argon between E_1 and *S* is ionised and later between E_2 and *S* so that a discharge starts warming up the interior which results in the mercury being gradually evaporated and the pressure within the lamp increases. In a few minutes the colour of the glow within the tube changes from the red of the neon to the greenish blue of mercury. The lamp in addition to intrinsic green violet gives a copious supply of ultraviolet light. The spectrum is found to consist of a number of lines which are used for the study of several phenomena.

The form (*b*) consists of an evacuated tube ending in two bulbs containing mercury. Two metallic electrodes are dipping in them and can be connected to the mains through a variable resistance *R*. By slightly tilting the tube an arc is struck which emits light in ultraviolet.

A laboratory form of a sodium vapour lamp. *A* is a U-shaped tube made of some special glass not blackened by sodium vapour. It has two oxide coated electrodes E_1 and E_2 sealed into its ends and contains sodium and neon. The electrodes are connected to the secondary of a transformer giving a sufficiently high voltage to start the discharge in the neon. The inner assembly is surrounded by a vacuum flask F to avoid heat losses and soon the temperature of A rises so that sodium volatilises and carries the discharge, being more easily ionised than neon atoms. In a few minutes the working temperature of about 300°C is reached and the red colour of the neon gives place to intense yellow D_1 and D_2 sodium lines.

The Spark

If two metal electrodes are connected to the terminals of the secondary of an induction coil or that of a high voltage transformer, a series of sparks may be produced across the gap provided it is small. The intensity of the spark can be increased by joining a condenser across the gap and the spark. This provides an intense source of radiation. The spectrum though characteristic of the metal forming the electrodes is different from that obtained by an arc. In the arc we have low voltages and high currents, whereas in the spark the voltages are as high as many thousands and the currents are small.

The Vacuum or Discharge Tube

This is a common laboratory source for the spectral analysis of light emitted by a gas through which a discharge is passed at low pressure. Two simple forms of discharge tubes containing the gas under examination. It consists of two wider portions A and B joined together by a capillary tube C.

The aluminium electrodes are fitted into the wider tubes and are held in position by platinum wires sealed into the tubes. By connecting the electrodes to an induction coil an electric discharge is passed through the gas in the tube and a brilliant luminosity results. Tubes of various shapes may be used to suit the convenience of the working conditions but the principle is the same.

Besides the sources mentioned above there are several others used in the present-day technique for the examination of spectra; for instance, the fluorescent tube, the electronic bombardment and the electrodeless discharge obtained by means of Tesla coil which produce high frequency electrical currents If a partially exhausted tube without electrodes is placed inside the Tesla coil, it emits a brilliant glow which is characteristic of the gas at low pressures. Various other types of arcs such as pulsating direct current arc, the high voltage alternating current arc are also available for the purpose and are usefully employed in spectroscope.

Emission and Absorption Spectra

The spectra may be grouped into two principal classes known as emission spectra and absorption spectra. The former are due to the light emitted from any of the sources described in the preceding section, while the latter are due to the absorption of light energy by material media interposed in the path of light. This absorption may be *general* when the light of all wavelengths within the visible region of the spectrum is absorbed during its passage through a material medium or *selective* in which case some colours are absorbed and some transmitted such as in coloured glasses and dyes.

Emission Spectra

It is usual to distinguish between two main types: *i*) Continuous spectra, and *ii*) discontinuous spectra.

Continuous spectra contains all the colours emerging from one to the other. They are emitted by any solid body when white hot. A dense luminous gas flame gives a continuous spectrum due to the presence of the incandescent carbon particles in it. The continuous spectrum is caused by the molecules in the solid being very close together and packed up so that collisions are frequently and constantly occurring and therefore all possible vibrations and wavelengths are emitted.

Discontinuous spectra may be classified into two groups: (*i*) line spectra, and (*ii*) band spectra.

(i) *Line Spectra:* When light is emitted by a substance in the atomic state such as sodium vapour in a sodium lamp or a glowing gas and is examined by a prism spectrometer or grating, it is found to consist of a number of isolated bright lines of different wavelengths separated by dark gaps. Such a spectrum is called a line spectrum. Salts of sodium, potassium and calcium in the Bunsen flame give line spectra Similarly, if an electric discharge is passed through hydrogen or helium at low pressure, line spectra are produced. Such spectra are generally emitted by the substances in the *atomic* state and are *characteristic* of them. Very often these lines are regularly spaced and form what we call a *series*.

It may also happen that the lines may be quite irregularly spaced.

(ii) *Band Spectra:* These spectra consist of a number of bright bands the intensity of which varies from one end to the other. In fact on examination by spectroscopes of high resolving power, it is found that each band consists of fine lines which are very close to each other at the sharp end of the band and are gradually more and more distant at the diffuse end. These spectra have a fluted appearance and are very often known as fluted spectra. Such spectra are emitted by the substances in the molecular state, when the heat is not too strong to break the molecules into constituent atoms. The carbon compounds mostly reveal this type of spectrum. In the discontinuous spectrum, the molecules and the atoms are on the average much farther apart and do not interfere with each other's radiation.

Although, in general, line spectra originate in atoms only and band spectra in molecules, yet the same substance may give a variety of spectra when excited to luminosity by different means. This is the case with several substances such as sodium, nitrogen, etc. With sodium in Bunsen flame two close yellow lines are observed while the photograph of sodium arc reveals many spectral lines. Again if the spark of sodium is examined by a spectroscope certain other lines are found to be present. We then have three different kinds of spectra, all characteristic of the sodium with some common lines but no two spectra are exactly alike.

Spectrum Analysis

The work of Bunsen and Kirch off revealed many interesting facts about spectra, most important being that the emission spectra particularly in the vaporous condition depend entirely on the nature of the substance. Therefore, the spectroscope examination of the light emitted by a substance enables us to find its composition. The method is *rapid* and *sensitive,* for even a small trace of the substance in the Bunsen flame is sufficient to give the characteristic lines of the substance. This method of identifying a substance is called spectrum analysis. The spectra of all substances have been studied and tabulated so that if we want to find out the composition of an unknown substance the arc spectrum of the substance is examined in a spectroscope and compared with the known spectrum. This method has made possible the discovery of a number of elements available in small quantities such as caesium, rubidium, thallium, helium, etc.

Besides this practical application of the spectra of the elements, a good deal of information is obtained from them with regard to the structure of the atoms. The spectra are in fact the language of the atoms which has been interpreted now and most of our present information about atom building has been obtained from their study.

Spectra of Various Substances

We have now studied spectroscopically the spectra of various elements and find that they differ greatly in complexity. Thus structure of the *H*-spectrum is the simplest as it gives four bright lines, in the visible region, one in the extreme red, one in the blue, one in the blue-violet and one in the violet. These lines are known as

$$H_\alpha, H_\beta, H_\gamma, \text{ and } H_\delta.$$

Besides we know now, that the hydrogen spectrum extends in the ultraviolet and infrared regions giving several other series of lines. The next in the order of simplicity are the spectra of the alkali metals, lithium, sodium, potassium, etc. Their spectra are strikingly similar to each other. We find that most of the lines are doublets as in the case of sodium where we have D_1 and D_2 lines very close altogether with wavelengths 5896 and 5890 Å.

By means of the spectroscope the spectra of most of the elements have been investigated and in the case of rare earths three close lines have been discovered. The spectra of some substances are very complex. Iron, for instance, gives as many as 5300 lines which have all been measured accurately.

The spectra of elements somewhere in the middle of the periodic table contain in general a large number of lines apparently distributed at random but in most cases they are regularly spaced and arranged in some systematic order.

Optical Lenses

Lens' Optical Centre

Let C_1 and C_2, be centres of curvature the of faces of a lens placed in a uniform medium and draw two parallel radii C_1Q, C_2R, then the tangents at Q and R will obviously be parallel to each other. Hence, a ray PQ which on refraction traverses the path QR will emerge parallel to PQ as the two refractions which it suffered at Q and R occurred at opposite parallel surfaces of a refracting medium. The point O where the ray QR cuts the axis is called the optical centre of the lens.

It is required to locate the position of O. By similar triangles C_1OQ and C_2OR, we have

$$\frac{OC_1}{OC_2} = \frac{QC_1}{RC_2} = \frac{O_1C_1}{O_2C_2} = \frac{r_1}{r_2}.$$

Also since $$\frac{O_1C_1}{O_2C_2} = \frac{OC_1}{OC_2}$$

$$= \frac{O_1C_1 - OC_1}{O_2C_2 - OC_2} = \frac{O_1O}{O_2O}$$

$$O_1O : O_2O = r_1 : r_2. \quad \text{... (17)}$$

That is, for a given lens, the position of O is fixed. Hence, any ray which emerges from the lens in a direction parallel to the incident ray passes through point O.

It is through this point that an undeviated ray is drawn in all cases. Next produce the rays PQ and RS forwards and backwards , respectively to cut the axis in N_1, N_2, Then N_1, N_2 are the first and the second nodal points of the lens. They are conjugate to O with regard to the *first* and *second* retracting surfaces as a ray incident through N_1 passes through O after *first* refraction and on emergence passes through N_2 parallel to its original direction.

Hence, for a lens in a uniform medium the principal points (which coincided in this case with the nodal points) are the points conjugate to the optical centre with regard to the two refracting surfaces. By making use of equation (17), we can locate the positions of the nodal points.

Since $$\frac{O_1O}{O_2O} = \frac{r_2}{r_1}$$

and $$O_1O_2 = t, \text{ we get}$$

$$= O_2O - t = \frac{r_2}{r_1} O_1O - t,$$

i.e., $$O_1O = \frac{r_1 t}{r_2 - r_1}$$

If N_1 is regarded as the point object in air for the first surface

$$\frac{\mu}{v'} - \frac{1}{O_1N_1} = \frac{\mu - 1}{r_1},$$

where $$v' = -O_1O.$$

$$\therefore \quad \frac{\mu(r_2 - r_1)}{r_1 t} - \frac{1}{O_1N_1} = \frac{\mu - 1}{r_1},$$

or $$= \frac{\mu(r_2 - r_1)}{r_1 t} - \frac{\mu - 1}{r_1}$$

$$= \frac{\mu(r_1 - r_2)}{r_1 t} - \frac{\mu - 1}{r_1}$$

$$= \frac{\mu(r_1 - r_2) - t(\mu - 1)}{r_1 t}$$

$$= \frac{\mu(r_1 - r_2) + t(1 - \mu)}{r_1 t}$$

$\therefore$ $$O_1N_1 = \frac{r_1 t}{\mu(r_1 - r_2) + t(1 - \mu)} \qquad \text{... (18)}$$

Similarly, $$O_2N_2 = \frac{r_2 t}{\mu(r_1 - r_2) + t(1 - \mu)} \qquad \text{... (19)}$$

Positions of cardinal points in some special cases.

Sphere Surrounded by Air

(i) *Distances fo Principal Points:* The distance of the first principal point from the first vertex,

$$\alpha = O_1H_1 = \frac{f_1 t}{\mu(f_2 - f_1) - t}.$$

Now, $$f_1 = -\frac{r_1}{\mu - 1} = -\frac{r}{\mu - 1} \qquad \text{since } r_1 = r$$

$$f_2 = -\frac{r_2}{\mu - 1} = \frac{r}{\mu - 1} \qquad \text{since } r_2 = -r$$

and $t = 2r.$

$\therefore$ $$\alpha = \frac{-\frac{r}{\mu - 1} \times 2r}{\mu\left(\frac{r}{\mu - 1} + \frac{r}{\mu - 1}\right) - 2r} = \frac{\frac{2r^2}{\mu - 1}}{\mu \frac{2r}{\mu - 1} - 2r}$$

$$= \frac{2r^2}{\mu - 1} \times \frac{\mu - 1}{2\mu r - 2\mu r + 2r} = \frac{2r^2}{2r} = r.$$

The distance of the second principal point from the second vertex,

$$\beta = O_2H_2 = \frac{f_2 t}{\mu(f_2 - f_1) - t}$$

$$= -\frac{\dfrac{r}{\mu-1} \times 2r}{\mu\left(\dfrac{r}{\mu-1} + \dfrac{r}{\mu-1}\right) \times 2r} = \frac{\dfrac{2r^2}{\mu-1}}{\dfrac{2\mu r}{\mu-1} - 2r}$$

$$= -\frac{2r^2}{2r} = -r.$$

Thus, the two principal points of a sphere are coincident with the centre of the sphere. Since the sphere is surrounded with air, the nodal points also coincide with the centre and so is the case with the optical centre. Hence, all the points are coincident with the centre of the sphere.

(ii) *Focal Lengths:* The distance of the first principal focus from the first principal plane and that of the second principal focus from the second principal plane, *viz.*, H_1F_1 and H_2F_2 are each equal to *conventionally* $H_1\ F_1$ is negative and H_2F_2 is positive.

Now, $$F = -\frac{\mu f_1 f_2}{\mu(f_2 - f_1) - t}$$

$$= -\frac{\mu \times \dfrac{-r}{\mu-1} \times \dfrac{r}{\mu-1}}{\mu\left(\dfrac{r}{\mu-1} + \dfrac{r}{\mu-1}\right) - 2r} = \frac{\dfrac{\mu r^2}{(\mu-1)^2}}{\mu\left(\dfrac{2r}{\mu-1}\right) - 2r}$$

$$= \frac{\mu r^2}{(\mu-1)^2} \times \frac{\mu - 1}{2\mu r - 2\mu r + 2r}$$

$$= \frac{\mu r^2}{2r(\mu-1)} = \frac{\mu r}{2(\mu-1)}$$

The sphere acts as a converging refracting system and the equivalent thin lens would be a convergent lens of numerical focal length $\frac{\mu r}{2(\mu - 1)}$ placed at the centre of the sphere.

Plano-convex Lens Surrounded by Air

(i) Distances of principal points.

Here, $r_1 = \infty; \quad r_2 = -r,$

hence $$\frac{1}{f_1} = -\frac{\mu - 1}{r_1} = \frac{\mu - 1}{\infty} = 0$$

and $$\frac{1}{f_2} = -\frac{\mu - 1}{r_2} = \frac{\mu - 1}{r}.$$

Now, $$\alpha = O_1H_1 = -\frac{f_1 t}{\mu(f_2 - f_1) - t}$$

$$= \frac{-t}{\mu\left(\frac{f_2}{f_1} - 1\right) - \frac{t}{f_1}} = \frac{-t}{\mu\frac{f_2}{f_1} - \mu - \frac{t}{f_1}}$$

$$= +\frac{t}{\mu} \qquad \ldots \left[\text{since } \frac{1}{f_1} = 0\right]$$

and $$\beta = O_2H_2 = -\frac{f_2 t}{\mu(f_2 - f_1) - t}$$

$$= \frac{-t/f_1}{\frac{\mu}{f_1} - \frac{\mu}{f_2} - \frac{t}{f_1 f_2}} = 0 \ldots \left[\text{since } \frac{1}{f_1} = 0\right]$$

As the lens is surrounded by air, the principal points and the nodal points are coincident.

(ii) Focal lengths

$$H_2F_2 = -H_1F_1 = F$$

Now,
$$F = -\frac{\mu f_1 f_2}{\mu(f_2 - f_1) - t}$$

$$= -\frac{\mu}{\frac{\mu}{f_1} - \frac{\mu}{f_2} - \frac{t}{f_1 f_2}}$$

$$= -\frac{\mu}{\frac{-\mu}{f_2}} = f_2 \quad \ldots \left[\text{since } \frac{1}{f_1} = 0\right]$$

or
$$F = f_2 = \frac{r}{\mu - 1}$$

If convex side faces the incident beam, $\alpha = 0$,

$$\beta = -\frac{t}{\mu}$$

and
$$F = \frac{r}{\mu - 1}$$

Note that in the case of a hemispherical lens, $t = r$,

$$\alpha = \frac{r}{\mu}, \quad \beta = 0 \text{ and } F = \frac{r}{\mu - 1}.$$

This means that for a hemispherical lens the distances of the principal foci form the principal points are the same as in the case of a plano-convex lens. If for such a lens $\mu = 3/2$ and PQ is the position of the object, then for refraction at first surface

$$O_1Q_1 = \mu O_1Q = \frac{3}{2}O_1Q$$

$$\alpha = O_1H_1 = \frac{r}{\mu} = \frac{r}{3/2} = \frac{2}{3}r; \quad \beta = O_2H_2 = 0,$$

i.e., H_2 coincides with O_2,

$$F = \frac{r}{\mu - 1} = \frac{r}{3/2 - 1} = \frac{r}{1/2} = 2r,$$

$\therefore \qquad H_1F_1 = -2r.$

Now, $\qquad O_1F_1 = H_1F_1 - H_1O_1$

$$= H_1F_1 + O_1H_1 = -2r + \frac{2}{3}r = -\frac{4}{3}r,$$

that is, the first principal focus is at distance $4r/3$ to the left of O.

Also $\qquad H_2F_2 = O_2F_2$, since H_2 and F_2 are coincident.

Since $\qquad H_2F_3 = F = 2r,$

$\therefore \qquad O_2F_2 = 2r.$

Example: *A glass sphere ($\mu = 1.50$) has a radius 1.5 cm. Calculate the positions of the cardinal points and also the position of the image when the object is 1.5 cm from the nearest point of the sphere assuming paraxial rays.*

Here, $\qquad r_1 = +1.5$ cm, $\qquad r_2 = -1.5$ cm,
$t = 3$ cm; $\qquad u = 1.5$ cm

$$\therefore \qquad f_1 = -\frac{r_1}{\mu - 1} = -\frac{+1.5}{1.5 - 1} = -3 \text{ cm}$$

$$f_2 = -\frac{r_2}{\mu - 1} = -\frac{-1.5}{1.5 - 1} = +3 \text{ cm}$$

$$\alpha = O_1H_1 = -\frac{f_1 t}{\mu(f_2 - f_1) - t} = -\frac{-3 \times 3}{1.5(3+3) - 3}$$

$$= +9/6 = +1.5 \text{ cm}$$

$$\beta = O_2H_2 = -\frac{f_2 t}{\mu(f_2 - f_1) - t} = -\frac{3 \times 3}{1.5(3+3) - 3}$$

$$= -9/6 = -1.5 \text{ cm}$$

Thus, the principal points *coincide* with the centre of the sphere. Since the sphere is surrounded with air, the nodal points also coincide with the centre and so is the case with the optical centre.

$$F = H_2F_2 = -\frac{\mu f_1 f_2}{\mu(f_2 - f_1) - t}$$

$$= -\frac{1.5 \times (-3) \times 3}{1.5(3+3)-3}$$

$$= \frac{13.5}{6} = 2.25 \text{ cm.}$$

To find the position of the image use the relation.

$$\frac{1}{V} - \frac{1}{U} = \frac{1}{F}$$

Here, $U = u - \alpha = -1.5 - 1.5 = -3.0$

$$\therefore \qquad \frac{1}{V} - \frac{1}{3} = \frac{1}{2.25}$$

or $$\frac{1}{V} = \frac{1}{2.25} - \frac{1}{3} = \frac{25}{225} = \frac{1}{9}$$

or $V = 9$ cm

Since $v - \beta = V, v = V + \beta = 9 - 1.5 = 7.5$ cm,

that is, image is on the other side at a distance 75 cm from the nearest surface.

Example: *A plano-convex lens 3 cm thick is made of glass of refractive index 150. If the first surf ace has a radius 3 cm, find (a) the focal length of the lens, (b) the power P and (c) the distances from the first vertex to the two principal planes.*

Here $r_1 = +3$ cm ; $r_2 = \infty$,
$t = 3$ cm, $\mu = 1.5$.

$$\therefore \qquad \frac{1}{f_1} = -\frac{\mu - 1}{r_1} = -\frac{1.5-1}{+3} = -\frac{0.5}{3} = -\frac{1}{6}.$$

$$\frac{1}{f_2} = -\frac{\mu - 1}{r_2} = -\frac{1.5-1}{\infty} = 0$$

(*a*) $$F = -\frac{\mu f_1 f_2}{\mu(f_2 - f_1) - t} = -\frac{\mu}{\frac{\mu}{f_1} - \frac{\mu}{f_2} - \frac{t}{f_1 f_2}}$$

$$= -\frac{1.5}{-\frac{1.5}{6}} = +6 \text{ cm.}$$

(*b*) Power $P = 100/f = +100/6 = +16.67$ D.

(*c*) $$\alpha = O_1H_1 = -\frac{f_1 t}{\mu(f_2 - f_1) - t}$$

$$= -\frac{t/f_2}{\left(\frac{\mu}{f_1} - \frac{\mu}{f_2}\right) - \frac{t}{f_1 f_2}} = 0 \qquad \ldots\left[\because \frac{1}{f_2} = 0\right]$$

$$\beta = O_2H_2 = -\frac{f_2 t}{\mu(f_2 - f_1) - t}$$

$$= -\frac{t/f_1}{\left(\frac{\mu}{f_1} - \frac{\mu}{f_2}\right) - \frac{t}{f_1 f_2}} = 0$$

$$= -\frac{3 \times (1/6)}{-1.5/6} = \frac{+1/2}{-3/12}$$

$$= -\frac{1}{2} \times \frac{12}{3} = -2 \text{ cm}$$

$$O_1H_1 = \alpha = 0;\ O_1H_2 = O_1O_2 - H_2O_2$$

$$= O_1O_2 + O_2H_2 = 3 - 2 = +1 \text{ cm}$$

That is distances from the first vertex to the first and the second principal points are zero and 1 cm , respectively.

For Ready Reference

(a) For a thick lens of thickness *t*, refractive index m and with f_1 and f_2 as the first focal distance of the first surface and second focal distance of second surface , respectively,

$$F = -\frac{\mu f_1 f_2}{\mu(f_2 - f_1) - t}$$

$$\beta = -\frac{f_2 t}{\mu(f_2 - f_1) - t}$$

and
$$\alpha = -\frac{f_1 t}{\mu(f_2 - f_1) - t}.$$

(b) For Two thin lenses of focal lengths, f_1 and f_2 separated by a distance t

$$F = +\frac{f_1 f_2}{f_2 + f_1 - t}$$

$$\beta = -\frac{f_2 t}{f_1 + f_2 - t}$$

and
$$\alpha = +\frac{f_1 t}{f_1 + f_2 - t}.$$

In calculating F, β and α in either case, the usual sign conventions to the given quantities must be applied.

Lens that are Thick

In dealing with thin lenses we assumed that the thickness of a lens is small and hence negligible as compared to the focal length, the object and image distances, which are always measured from the optical centre of the lens. But, when the lens is sufficiently thick this procedure of measuring distances is not applicable and we have to deal with it like a system of coaxial lenses treated as a unit which may or may not be in contact. Such a lens or a system of lenses must be treated as a thick lens.

To find the position of image in this case we may apply the equation of a single refracting surface successively to each surface. But with a large number of surfaces the procedure is laborious and as such, the treatment suggested by C.F. Gauss is followed. Having studied the properties of thick lenses as a whole he has shown that simple formulae as for thin lenses may be applied as usual provided, however, the distances in all cases are measured from two points fixed relatively to the lens called principal points. Besides these two principal points, there are two principal foci and two nodal points as in coaxial lens systems. When the medium on both sides of the thick lens is the same, the nodal points are coincident with the principal points. These six points, already termed as cardinal points, have as well an important bearing on the theory of thick lenses.

Principal Foci and Principal Points: Rays proceeding from the point F_1 emerge parallel to the axis after refraction from the lens and the parallel incident rays are brought to focus at F_2. F_1 and F_2 are called the first and second *principal foci* , respectively. The transverse planes through the points of intersection of the incident and the emergent rays are called the first and second *principal planes* and the points H_1 and H_2 where they cut the principal axis are called *principal points.* These points have the same significance as discussed already, with reference to the coaxial system of lenses. The positions of the points F_1, F_2, H_1 and H_2 are fixed for a lens. If the medium is the same on both sides of the lens, the first and second focal lengths ϕ_1 and ϕ_2 will be numerically equal.

Note that the focal lengths are measured from the principal foci to their respective principal points H_1 and H_2 and not to their respective vertices.

In general the principal points and the principal foci are not located symmetrically with respect to the lens but are at different distances from the vertices. The dotted lines indicate the approximate positions of the principal planes in the case of a few lenses of the standard shapes.

Now, we shall discuss the behaviour of a thick lens bounded on both sides by the same (less dense) medium, for instance, a glass lens placed in air.

Thick Lens Formulae: Assuming that the medium on both sides of the lens is air, we shall find the positions of the image, the principal points, the principal foci, etc., for a thick lens by the same general method as used for combination of thin liases treating the image formed by the first surface as the object for the second.

Let PQ, be an object placed at a distance u from the first surface of a lens of thickness t, refractive index μ, with faces of radii of curvature r_1 and r_2, for refraction at the first surface where v' is the distance of the image from the surface and f_1 is the first focal distance of this surface.

$$\frac{\mu}{v'} - \frac{1}{u} = \frac{\mu - 1}{r_1} = -\frac{1}{f_1}, \qquad \text{... (1)}$$

where v' is the distance of the image from the surface and f_1 is the first focal distance of this surface.

Since the image formed by the first surface becomes the object for the second surface, we have for refraction at the second surface $u' = v' - t$, due regard being paid to the sign of v'.

Since the rays are now going from a denser to a rarer medium, we have by the same relation.

$$\frac{\frac{1}{\mu}}{\mu} - \frac{1}{v' - t} = \frac{\frac{1}{\mu} - 1}{r_2} \qquad \text{... (2)}$$

or
$$\frac{1}{v} - \frac{\mu}{v' - t} = \frac{1 - \mu}{r_2} = -\frac{\mu - 1}{r_2} = \frac{1}{f_2} \qquad \text{... (2a)}$$

where v is the distance of the final image from the second surface and f_2 is the second focal distance for the second surface. It is the distance of the point from the second surface where a beam of parallel rays traversing the more dense medium incident on the second surface converges after emergence from the surface. It is the value of v obtained by putting $v' = \infty$ in equation (2).

By combining relations (1) and (2*a*) and eliminating v', we obtain by algebra,

$$\frac{1}{v - \beta} - \frac{1}{u - \alpha} = \frac{1}{F}$$

where F is the focal length of the thick lens, α and β are the distances of the first principal plane and the second principal plane from the first surface and the second surface of the lens, respectively.

Here, $$F = -\frac{\mu f_1 f_2}{\mu(f_2 - f_1) - t} \qquad \dots (4)$$

$$\beta = -\frac{f_2 t}{\mu(f_2 - f_1) - t} \qquad \dots (5)$$

and $$\alpha = -\frac{f_1 t}{\mu(f_2 - f_1) - t}$$

If we measure V from a point at a distance β from the second surface and U from a point at a distance α from the first surface, then we may write

$$V = v - \beta$$

and $$U = u - \alpha$$

Equation (3) thus becomes

$$\frac{1}{V} - \frac{1}{U} = \frac{1}{F}, \qquad \dots (7)$$

a formula identical with that used for a thin lens. U and V are known as the reduced object distance and the reduced image distance , respectively.

The point at a distance α from the first surface of the lens is called the first principal point of the lens and the point at a distance β from the second surface of the lens is termed the second principal point. Planes drawn perpendicular to the axis through the principal points are called the *principal planes* of the lens. Knowing the constants of the lens the values of β and α are calculated from equations (5) and (6) and hence the positions of principal points are determined.

Having thus located the principal points, the principal foci can be easily found. *The first principal focus is at a distance F to the left of the first principal point and the second principal focus is at the same distance to the right of the second principal point.* The first and second focal lengths of a lens surrounded by a uniform medium are, therefore, equal in magnitude but opposite in sign.

Since α, β and F depend only on the constants r_1, r_2, t and μ, their values in any given problem can be readily found. Hence, equation (3) provides us with the desired means of locating the image formed by a thick lens.

The positions of the principal points H_1, H_2, the principal foci F_1, F_2 and the course of rays forming the image $P'Q'$ of an object PQ placed at a distance u from the first surface of a thick lens. For tracing the course of rays.

The relations (3) to (6) can be obtained as below:

For refraction at the first and the second surface of a thick lens, we have

$$\frac{\mu}{v'} - \frac{1}{u} = -\frac{1}{f_1} \qquad \text{... } (i)$$

and
$$\frac{1}{v} - \frac{\mu}{v'-t} = \frac{1}{f_2}. \qquad \text{... } (ii)$$

From (*i*),
$$\frac{\mu}{v'} = \frac{1}{u} - \frac{1}{f_1} = \frac{f_1 - u}{uf_1}$$

∴
$$v' = \frac{\mu u f_1}{f_1 - u}. \qquad \text{... } (iii)$$

From (*ii*),
$$\frac{\mu}{v'-t} = \frac{1}{v} - \frac{1}{f_2} = \frac{f_2 - v}{vf_2}.$$

∴
$$v' - t = \frac{\mu v f_2}{f_2 - v},$$

or
$$v' = t + \frac{\mu v f_2}{f_2 - v}, \qquad \text{... } (iv)$$

Equating the values of v' in (*iii*) and (*iv*), we have

$$\frac{\mu v f_1}{f_1 - u} = t + \frac{\mu v f_2}{f_2 - v}$$

Multiplying throughout by $(f_1 - u)(f_2 - v)$, we get

$$\mu u f_1 (f_2 - v) = t (f_2 - v)(f_1 - u) + \mu v f_2 (f_1 - u),$$

or $\mu u f_1 (f_2 - v) - t (f_2 - v)(f_1 - u) - \mu v f_2 (f_1 - u) = 0$

Opening the brackets and collecting the terms in uv, u and v.

$$\mu u f_1 f_2 - \mu u v f_1 - t f_1 f_2 + t u f_2 + t v f_1 - t u v$$
$$- \mu v f_1 f_2 + \mu u v f_2 = 0,$$

or
$$\mu u v f_2 - \mu u v f_1 - t u v + \mu u f_1 f_2 + t u f_2$$
$$- \mu v f_1 f_2 + t v f_1 - t f_2 f_1 = 0,$$

or
$$uv[\mu(f_2 - f_1) - t] + u[\mu f_1 f_2 + t f_2]$$
$$- v[\mu f_1 f_2 - t f_1] - t f_1 f_2 = 0$$

Dividing throughout by $\mu(f_2 - f_1) - t$, we have

$$uv + \frac{u(\mu f_1 f_2 + t f_2)}{\mu(f_2 - f_1) - t} - \frac{v(\mu f_1 f_2 + t f_2)}{\mu(f_2 - f_1) - t} - \frac{t f_1 f_2}{\mu(f_2 - f_1) - t} = 0. \quad \text{... } (v)$$

the equation can be put in the form

$$\frac{1}{v - \beta} - \frac{1}{u - \alpha} = \frac{1}{F} \quad \text{... } (vi)$$

Multiplying (vi) throughout by $F(u - \alpha)(v - \beta)$ and collecting terms in uv, u and v, we obtain

$$uv - u(F + \beta) + v(F - \alpha) - F(\beta - \alpha) + \alpha\beta = 0 \quad \text{... } (vii)$$

In order that (v) and (vii) should be identical , we must have

$$- F - \beta = \frac{\mu f_1 f_2 + t f_2}{\mu(f_2 - f_1) - t} \quad \text{... } (viii)$$

$$- F + \alpha = \frac{\mu f_1 f_2 + t f_1}{\mu(f_2 - f_1) - t} \quad \text{... } (ix)$$

and
$$F(\beta - \alpha) - \alpha\beta = \frac{t f_1 f_2}{\mu(f_2 - f_1) - t} \quad \text{... } (x)$$

Thus (*viii*), (*ix*) and (*x*) are three simultaneous equation to find the three unknown α, β and F.

To solve these, first multiply the corresponding sides of (*viii*) and (*ix*). This gives us

$$F^2 - F\alpha + F\beta - \alpha\beta = \frac{\mu f_1 f_2 + tf_2}{\mu(f_2 - f_1) - t} + \frac{\mu f_1 f_2 + tf_1}{\mu(f_2 - f_1) - t} \quad \text{... } (xi)$$

Subtracting (*x*) from (*xi*), we get

$$F^2 = \frac{(\mu f_1 f_2)^2 - \mu f_1^2 f_2 t + \mu f_1 f_2^2 t - t^2 f_1 f_2}{[\mu(f_2 - f_1) - t]^2} - \frac{tf_1 f_2}{\mu(f_2 - f_1) - t}$$

$$= \frac{(\mu f_1 f_2)^2 - \mu f_1^2 f_2 t + \mu f_1 f_2^2 t - t^2 f_1 f_2 - tf_1 f_2 [\mu(f_2 - f_1) - t]}{[\mu(f_2 - f_1) - t]^2}$$

$$= \frac{(\mu f_1 f_2)^2 + tf_1 f_2 [\mu(f_2 - f_1) - t] - tf_1 f_2 [\mu(f_2 - f_1) - t]}{[\mu(f_2 - f_1) - t]^2}$$

$$= \frac{(\mu f_1 f_2)^2}{[\mu(f_2 - f_1) - t]^2}$$

$$\therefore \quad F = \pm \frac{\mu f_1 f_2}{\mu(f_2 - f_1) - t} \quad \text{... } (xii)$$

In order to find out whether the positive or the negative sign is to be retained it should be noticed that it must be true for all values of t. For a thin lens $t = 0$, and obviously f_1 and f_2 are conjugate focal distances and α and β are both zero.

Substituting $\alpha = 0$, $u = f_1$ and $\beta = 0$, $v = f_2$ in (*vi*), we have

$$\frac{1}{f_2} - \frac{1}{f_1} = \frac{1}{F}$$

$$\therefore \quad F = -\frac{f_1 f_2}{f_2 - f_1}.$$

Substituting $t = 0$ in (*xii*), we see that we obtain value of F just found, provided that we take the negative sign. Hence,

$$F = \frac{\mu f_1 f_2}{\mu(f_2 - f_1) - t} \qquad \text{... } (xiii)$$

Adding (*xiii*) and (*viii*), we have

$$-\beta = \frac{\mu f_1 f_2 + t f_2}{\mu(f_2 - f_1) - t} - \frac{\mu f_1 f_2}{\mu(f_2 - f_1) - t}$$

$$= \frac{t f_2}{\mu(f_2 - f_1) - t}$$

or

$$\beta = \frac{f_2 t}{\mu(f_2 - f_1) - t} \qquad \text{... } (xiv)$$

Adding (*xiii*) and (*ix*), we get

$$\alpha = \frac{\mu f_1 f_2 - t f_1}{\mu(f_2 - f_1) - t} - \frac{\mu f_1 f_2}{\mu(f_2 - f_1) - t}$$

$$= -\frac{f_1 t}{\mu(f_2 - f_1) - t} \qquad \text{... } (xv)$$

Hence,

$$\alpha f_2 = \beta f_1, \qquad \text{... } (xvi)$$

an important relation which is useful in the theory of thick lenses.

Power of a Thick Lens: The power of a thick lens is given by

$$P = \frac{1}{F} = -\frac{\mu(f_2 - f_1) t}{\mu f_1 f_2}$$

$$= -\frac{1}{f_1} + \frac{1}{f_2} + \frac{t}{\mu f_1 f_2}$$

$$= \frac{\mu - 1}{r_1} - \frac{\mu - 1}{r_2} - \frac{t}{\mu}\left(\frac{\mu - 1}{r_1}\right)\left(-\frac{\mu - 1}{r_2}\right)$$

$$= (\mu - 1)\left(\frac{1}{r_1} - \frac{1}{r_2}\right) + \frac{t(\mu - 1)^2}{\mu r_1 r_2} \qquad \text{... } (8)$$

The power of a thick lens is also obviously given by

$$P = P_1 + P_2 - \frac{t}{\mu} P_1P_2,$$

where P_1 and P_2 are the powers of the two surfaces , respectively.

Graphical Construction of the Image of an Object on the Axis of a Thick Lens: If F be the focal length of a thick lens and the object and the image distances be measured from the first and second principal planes , respectively, the usual graphical method of locating the images for thin lenses will apply to thick lenses as well.

PQ is an object, the position of whose image we wish to find graphically. First calculate from the given data the values of F, α and β. Then locate on the axis the principal points H_1 and H_2 by measuring from the first vertex $O_1H_1 = \alpha$, to the right if α is positive and to the left if it is negative, and from the second vertex $O_2H_2 = \beta$ in the same way while the principal planes are drawn though H_1 and H_2. The points F_1 and F_2, the first and the second principal foci are then located by taking $H_1F_1 = -F$ *and* $H_2F_2 = -F$. For the purposes of graphical construction of the image, the lens may be regarded as-replaced by its two principal planes and the properties of these planes mentioned already with reference to coaxial systems are taken into consideration.

It will be seen that ray 1 from P drawn parallel to the optic axis meets the first principal plane at M and the second at N at the same height and then passes through F_2, the second principal focus of the lens.

Similarly, ray 2 passing through F_1, the first principal focus of the lens meets the first principal plane at M' and is then rendered parallel to the axis. The point of intersection of these two rays gives, the required position of P', the image of P.

From P' drop $P'Q'$ perpendicular to the axis meeting it in Q, the $P'Q'$ is the image of PQ.

It may be noted that here all the rays in the region between two principal planes are drawn parallel to the axis since by definition these planes are conjugate ones, for which the magnification is unity and has a positive sign. It means that a small object on the axis in one of these planes has an equal-sized, erect image in the other. These planes are, therefore, known as *unit planes.*

Example: *Calculate the positions of the principal points and principal foci of a thick double convex lens, the radii of whose surfaces are 4 cm and 3 cm, the thickness of the lens being 2.5 cm and the refractive index 1.50.*

Here, $r_1 = +4$ cm; $r_2 = -3$ cm, $t = 2.5$ cm, $\mu = 1.5$.

$$\therefore \quad f_1 = -\frac{r_1}{\mu - 1} = -\frac{+4}{1.5-1} = -8 \text{ cm.}$$

$$f_2 = -\frac{r_2}{\mu - 1} = -\frac{-3}{1.5-1} = +6 \text{ cm}$$

$$F = -\frac{\mu f_1 f_2}{\mu(f_2 - f_1) - t} = -\frac{1.5 \times (-8) \times (+6)}{1.5[6-(-8)] - 2.5}$$

$$= +\frac{1.5 \times 48}{18.5} = +\frac{144}{37} = +3.89 \text{ cm approx.}$$

$$\alpha = -\frac{f_1 t}{\mu(f_2 - f_1) - t}$$

$$= \frac{-8 \times 2.5}{1.5(6+8) - 2.5} = +1.08 \text{ cm.}$$

$$\beta = \frac{f_2 t}{\mu(f_2 - f_1) - t}$$

$$= -\frac{+6 \times 2.5}{1.5(6+8) - 2.5} = -\frac{15}{18.5} = -0.81 \text{ cm;}$$

$$\therefore \quad O_1H_1 = \alpha = +1.08 \text{ cm;}$$

$$O_2H_2 = \beta = -0.81 \text{ cm};$$

$$H_1F_1 = F = 3.89 \text{ cm};$$

$$\therefore \quad O_1F_1 = H_1F_1 - O_1H_1$$

$$= 3.89 - 1.08 = 2.81 \text{ cm}.$$

$$H_2F_2 = F = 3.89 \text{ cm};$$

$$\therefore \quad O_2F_2 = H_2F_2 - O_2H_2$$

$$= 3.89 - 0.81 = 3.08 \text{ cm}.$$

Example: *With the following data for a thick lens, calculate* f_1, f_2, F, α *and* β *and also the distances of the principal points and principal foci from the first vertex of the lens.*

$r_1 = +3$ cm; $t = 2$ cm; $r_2 = -5$ cm; $\mu = 1.50$.

Here,

$$f_1 = -\frac{r_1}{\mu - 1} = -\frac{+3}{1.5 - 1}$$

$$= -6 \text{ cm}.$$

$$f_2 = -\frac{r_2}{\mu - 1} = -\frac{-5}{1.5 - 1} = +10 \text{ cm}.$$

$$F = -\frac{\mu f_1 f_2}{\mu (f_2 - f_1) - t}$$

$$= -\frac{1.50 \times (-6) \times (+10)}{1.50\left[+10 - (-6) - 2\right]} = 4.09 \text{ cm}.$$

$$\beta = -\frac{f_2 t}{\mu (f_2 - f_1) - t}$$

$$= -\frac{(+10) \times (+2)}{1.50 (+10 + 6) - 2} = -0.90 \text{ cm}.$$

$$\alpha = -\frac{f_1 t}{\mu (f_2 - f_1) - t}$$

$$= -\frac{(-6) \times (+2)}{1.5 (+10 + 6) - 2} = +0.54 \text{ cm}.$$

The distance of the first principal point from the first vertex, α = + 0.54 cm, that is towards its right.

The distance of the second principal point from the second vertex = – 0.91 cm, that is, towards its left.

∴ its distance from the first vertex = 2 – 0.91 = 1.09 cm.

The distance of the first principal focus from the first vertex = 4.09 – 0.54 = 3.55 cm and the distance of the second principal focus from the first vertex = 2 + (4.09 – 0.91) = 5.18 cm.

Example: *An object is placed 50 cm from a thick double convex lens, with faces of radii of curvature 15 cm, and 20 cm, thickness 15 cm and refractive index 1.50. Find the position of the image.*

Here, we have

r_1 = + 15 cm; r_2 = – 20 cm; t = 1.5 cm; μ = 1.50, u = – 50 cm

$$\therefore \quad f_1 = -\frac{r_1}{\mu - 1} = -\frac{+15}{1.50 - 1} = -30.0 \text{ cm.}$$

$$f_2 = -\frac{r_2}{\mu - 1} = -\frac{-20}{1.50 - 1} = +40 \text{ cm.}$$

$$F = -\frac{\mu f_1 f_2}{\mu (f_2 - f_1) - t}$$

$$= -\frac{1.50 \times (-30) \times (+40)}{1.50 [+40 + 30] - 1.5} = +17.4 \text{ cm.}$$

$$\alpha = -\frac{f_1 t}{\mu (f_2 - f_1) - t}$$

$$= -\frac{(-30) \times (+1.5)}{1.50 (+40 + 30) - 1.5} = +0.43 \text{ cm.}$$

$$\beta = -\frac{f_2 t}{\mu (f_2 - f_1) - t}$$

$$= - \frac{(+40) \times (+1.5)}{1.50\,(+40 + 30) - 1.5} = - 0.58 \text{ cm.}$$

Applying the relation,

$$\frac{1}{v - \beta} - \frac{1}{u - \alpha} = \frac{1}{F}$$

and substituting the values of α, β and F, we have

$$\frac{1}{v - (-0.58)} - \frac{1}{-50 - (+0.43)} = +\frac{1}{17.4}$$

or

$$\frac{1}{v + 0.58} - \frac{1}{50.43} = +\frac{1}{17.4},$$

where v = 26.0 cm.

Refraction by a Thick Lens Separating Media of Different Refractive Indices: Let us suppose that the refractive index of the material of the lens is μ_2, that of the medium on the left μ_1 and that of the medium on the right μ_2.

Then for the first refraction,

$$\frac{\frac{\mu_2}{\mu_1}}{v'} - \frac{1}{u} = -\frac{1}{f_1}, \qquad \text{... }(a)$$

where
$$\frac{1}{f_1} = -\frac{\mu_2 - \mu_1}{\mu_1} \cdot \frac{1}{r_1}.$$

For the second refraction,

$$\frac{1}{v} - \frac{\frac{\mu_2}{\mu_3}}{v' - t} = \frac{1}{f_2} \qquad \text{... }(b)$$

where
$$\frac{1}{f_2} = \frac{\mu_3 - \mu_2}{\mu_3} \times \frac{1}{r_2}. \qquad \text{... }(b')$$

Changing (*a*) and (*b*) into the forms similar to (*i*) and (*ii*), we get

$$\frac{\mu_2}{v'} - \frac{1}{\frac{u}{\mu_1}} = -\frac{1}{\frac{f_1}{\mu_1}} \quad \text{... (c)}$$

and
$$\frac{1}{\frac{v}{\mu_2}} - \frac{\mu_2}{v' - t} = \frac{1}{\frac{f_2}{\mu_3}}. \quad \text{... (d)}$$

The terms μ_2, u/μ_1, v/μ_3, f_1/μ_1 and f_2/μ_3 of these equations correspond to the terms μ, u, v, f_1 and f_2 of equations (*i*) and (*ii*).

The solution of these equations is obtained by substituting: these values in equations (*vi*), (*xiii*), (*xiv*) and (*xv*).

Equation (*vi*) thus becomes

$$\frac{1}{\frac{v}{\mu_2} - \beta} - \frac{1}{\frac{u}{\mu_1} - \alpha} = \frac{1}{F},$$

or
$$\frac{\mu_2}{v - \mu_3\beta} - \frac{\mu_1}{u - \mu_1\alpha} = \frac{1}{F} \quad \text{... (e)}$$

Putting $U = u - \mu_1\alpha$ and $V = v - \mu_2\beta$, we may rewrite this equations as

$$\frac{\mu_3}{V} - \frac{\mu_1}{U} = \frac{1}{F} \quad \text{... (f)}$$

From Eqn. (*f*) it is obvious that keeping in mind the definitions of first and second focal lengths, the first focal length (ϕ_1) of the lens will be equal to $-\mu_1 F$ and the second focal length (ϕ_2) will be equal to $\mu_3 F$.

Then
$$\frac{\mu_3}{V} - \frac{\mu_1}{U} = -\frac{\mu_1}{\phi_1} = \frac{\mu_3}{\phi_3} \quad \text{... (g)}$$

We also find that the two focal lengths of the lens are *numerically* unequal being in the ratio of the refractive indices of the first and second medium.

Since $$\frac{\phi_1}{\phi_2} = \frac{-\mu_1 F}{\mu_3 F} = -\frac{\mu_1}{\mu_3} \qquad \text{... }(h)$$

The respective distance of the first and the second principal point from the first and the second surface of the lens will be equal to

$$\alpha' = \mu_1 \alpha$$

and $$\beta' = \mu_2 \beta.$$

Writing f_1/μ_1 for f_1, f_2/μ_3 for f_2 and μ in the values of α and β, we have

$$\alpha' = \mu_1 \alpha = -\frac{\mu_1 \dfrac{f_1 t}{\mu_1}}{\mu_2 \left(\dfrac{f_2}{\mu_2} - \dfrac{f_1}{\mu_1}\right) - t}$$

$$= -\frac{\mu_1 \mu_3 f_1 t}{\mu_2 (\mu_2 - \mu_2 f_1) - \mu_1 \mu_2 t}. \qquad \text{... (2)}$$

Similarly, $$\beta' = \mu_2 \beta = -\frac{\mu_1 \mu_3 f_2 t}{\mu_2 (\mu_2 f_2 - \mu_3 f_1) - \mu_1 \mu_3 t} \qquad \text{... (10)}$$

Obviously, $$f_2 \alpha' = f_1 \beta' \qquad \text{... (11)}$$

an important relation which is useful in the theory of thick lenses.

Also $$F = -\frac{\mu_2 \dfrac{f_1}{\mu_1} \times \dfrac{f_2}{\mu_3}}{\mu_2 \left(\dfrac{f_2}{\mu_3} - \dfrac{f_1}{\mu_1}\right) - t}$$

$$= -\frac{\mu_2 f_1 f_2}{\mu_2(\mu_1 f_2 - \mu_3 f_1) - \mu_1\mu_3 t} \qquad \text{... (12)}$$

$$\therefore \qquad \phi_2 = \mu_3 F$$

$$= -\frac{\mu_3\mu_2 f_1 f_2}{\mu_2(\mu_1 f_2 - \mu_2 f_1) - \mu_1\mu_3 t} \qquad \text{... (12a)}$$

and

$$\phi_1 = \mu_1 F$$

$$= -\frac{\mu_1\mu_2 f_1 f_2}{\mu_2(\mu_1 f_2 - \mu_3 f_1) - \mu_1\mu_3 t} \qquad \text{... (12b)}$$

With the help of equations (9) and (10) we can locate the positions of principal points. Since in this case the principal points do not coincide with the nodal points, the latter pair of points can be determined by combining with these results equations (16) and (16 *a*).

Having thus located the principal points, the principal foci can be easily found. The first principal focus is at a distance ϕ_1 to the left of the first principal point H_1, and the second principal focus is at a distance ϕ_2 to the right of the second principal point H_2.

It may be noted that the thin lens formulae, $\frac{1}{v} - \frac{1}{u} = \frac{1}{f}$

$$\frac{\mu_3}{v} - \frac{\mu_1}{u} = -\frac{\mu_1}{f_1} = \frac{\mu_3}{f_2}$$

and

$$\frac{f_1}{f_2} = \frac{\mu_1}{\mu_3},$$

connecting the conjugate points are applicable to a thick lens provided that the focal lengths, object and image distances are measured from the principal points and not from the vertices as in a thin lens.

The Principal Points of a Lens are Conjugate Foci: This can be proved in the case of a lens separating media of different refractive indices and then it evidently follows that it applies to a lens under any conditions.

Putting β' for $\mu_2\beta$ and α' for $\mu_1\alpha$ in equation *(e)*, we have

$$\frac{\mu_2}{v-\beta'} - \frac{\mu_1}{u-\alpha'} = \frac{1}{F}$$

Simplifying, $$v - \beta' = \frac{\mu_2 F \times (u-\alpha')}{\mu_1 F + (u-\alpha')} \qquad \text{...(13)}$$

When $u = \alpha$, we get $v - \beta' = 0$ or $v = \beta'$ for all values of F, μ_3, μ_1.

Hence, an object at a distance α' from the first surface of the lens (i e., an object at the first principal point) will produce an image at β' from the second surface of the lens (i.e., at the second principal point).

The Principal Planes of a Thick Lens are Planes of Unit Magnification: Let an object PQ, be placed on the axis of the lens and let F_1 be the first focal point of the first surface of the lens so that a ray of light PA starting from P and appearing to come from F_1 is rendered parallel to the axis along AB after refraction at the first surface. After refraction at the second surface the same ray will pass through F_2 the second focal point of the second surface of the lens. From the relations obtained already the position of the image P_1Q_1 can be located, that of P_1 being corresponding to the object point P. Drop perpendiculars AC and BD on the axis. When the slopes of the rays F_1A and BF_2 are small the feet of the perpendiculars will almost coincide with O_1 and O_3 the intersections of the axis with the faces of the lens.

Let $AC = BD = h$. $O_1Q = u$

and $O_2Q_1 = v$, so that

$$CQ = O_1Q = u \text{ approx.}$$

and $$DQ_1 = O_2Q_1 = v \text{ approx.}$$

If f_1 is the first focal distance of the first surface ($= -CF_1$) and f_2 the second focal distance of the second surface ($= DF_2$)

and if the heights of the object and the image are h_1 and h_2, respectively, then by similar Δs ACF_1 and PQF_1,

$$\frac{AC}{CF_1} = \frac{PQ}{QF_1} = \frac{PQ}{CF_1 - CQ},$$

or *conventionally*, $\dfrac{h}{-f_1} = \dfrac{h_1}{-f_1-(-u)} = \dfrac{h_1}{u-f_1}$

Also from Δs $P_1Q_1F_2$ and BDF_2

$$\frac{BD}{BF_2} = \frac{P_1O_1}{F_2Q_1} = \frac{P_1O_1}{DQ_1 - DF_2},$$

or *conventionally*, $\dfrac{h}{f_2} = \dfrac{-h_2}{v-f_2} = \dfrac{h_2}{f_2 - v}$.

Let m be the magnification produced by the lens, then

$$\mu = \frac{h_2}{h_1} = \frac{h(f_2 - v)}{f_2} \times \frac{f_1}{h(f_1 - u)}$$

$$= \frac{f_1(f_2 - v)}{f_2(f_1 - u)} = \frac{f_1f_2 - f_1v}{f_2f_1 - f_2u} \qquad \text{... (14)}$$

If now we suppose that the object lies in the first principal plane of the lens,

, $u = \alpha'$ as already proved, the corresponding image will be in the second principal plane, so that $v = \beta'$, hence

$$m = \frac{f_1f_2 - f_1\beta'}{f_1f_2 - f_2\alpha'} = +1 \qquad \text{... (15)}$$

since $\quad f_2\alpha' = f_1\beta' \quad$ by equation (11)

Consequently, any ray directed towards a point in the first principal plane at a distance h from the axis will give rise to a transmitted ray proceeding from a point in the second principal plane on the same side of the axis and at a distance h from it.

Nodal Points: Similar to the case of coaxial systems of lenses there is a pair of conjugate points on the axis of a thick lens called the first and second nodal points. They have the same properties as described earlier in the coaxial systems, *viz.*, that in the general case of a thick lens bounded by two different media *nodal points are defined as points on the axis such that an incident ray directed towards the first nodal point on emerging from the lens proceeds from the second nodal point in a direction parallel to that of the incident ray.* The nodal points and the principal points of a thick lens are also *coincident* when the medium on both sides of the lens is the *same.*

We shall now bring out the different significance of the nodal points and the principal points by considering a thick lens separating two media of different refractive indices in which case the two pairs of points are separate.

Let H_1 H_2 and N_1, N_2, be the principal and the nodal points , respectively of a thick lens and AF_1 an object in the first focal plane of the lens.

Then the ray AN_1 emerges from the lens in the parallel direction N_2D. Let the incident and the emergent rays intersect the principal planes in B and C , respectively. Then since principal planes are planes with unit magnification,

$$BH_1 = CH_2$$

and as AN_1 and N_2D are parallel,

$$N_1H_1 = N_2H_2.$$

∴ obviously $$N_1N_2 + H_1\ H_2. \qquad \text{... (16)}$$

Thus the distance between the nodal points is exactly equal to the distance between the principal points.

Now draw a ray AE parallel to the axis to cut the first principal plane in E. It will after refraction by the lens proceed from G a point in the second principal plane at an equal distance from the axis and pass through the second principal focus F_2. As A lies in the first focal plane of the lens, all rays proceeding from it will be rendered parallel after passing through the lens. Hence, GF_2 is parallel to N_2D and also to AN_1.

Thus the two Δs GF_2H_2 and AN_1F_1 are congruent and hence

$$H_2F_2 = N_1F_1 = H_1F_1 - H_1N_1,$$

or $$H_1N_1 = H_1F_1 = H_2F_2$$

Putting conventional signs

$$-H_1N_1 = -H_1F_1 - H_2F_2,$$

or $$H_1N_1 = H_1F_1 + H_2F_2 = \phi_1 + \phi_2, \quad \ldots (16a)$$

where ϕ_1 and ϕ_2 are the first and second focal lengths of the lens.

If the lens is surrounded with a *uniform medium*

$$\phi_1 = -\phi_2$$

and so $$H_1N_1 = 0$$

or the nodal points N_1 and N_2 coincide with the principal points H_1 and H_2 of the lens.

In other cases the positions of the nodal points are graphically determined by reversing the process above.

It may be noted that the knowledge of the six cardinal points namely, the two principal foci, the two principal points and the two nodal points is sufficient to solve all problems relating to a thick lens.

Lenses and Focal Planes

Focal Plane Theory

Planes at right angles to the axis through F_1 and F_2 are called principal focal planes. The property of the first principal focal plane is that if rays at incidence on a convergent lens, are diverging from a point P in the first focal plane, they emerge as a parallel beam, and if rays at incidence on a divergent lens, are converging to a point P in the first focal plane, they too, emerge as a parallel beam, the direction of the emergent pencil being given in both cases by the ray PO or OP which suffers no deviation.

Similarly, for the second principal focal plane if a parallel beam of light is incident on a convex lens, on emergence it will meet at a point in the second principal focal plane and in the case of a concave lens a parallel incident beam on emergence will appear to proceed from a point in the second principal focal plane.

Focal Length of a Thin Lens by Format's Principle: Let P be a point object on the principal axis of a thin convex lens and

Q its point image. Let us consider the two rays PAQ and PO_1 O_2Q going from P to Q.

By Format's principle ail the rays take a stationary or a maximum or a minimum time. If v_1 and v_2 be the velocities of light in the media of refractive indices μ_1 and μ_2 respectively, we have

$$\frac{PA + AQ}{v_1} = \frac{PO_1 + O_2Q}{v_1} + \frac{O_1O_2}{v_2}$$

or $$PA + AQ = PO_1 + O_2Q + \frac{v_1}{v_2} O_1O_2$$

But $$\frac{v_1}{v_2} = \frac{\mu_2}{\mu_1}$$

$$\therefore \mu_1 (PA + AQ) = \mu_1 (PO_1 + O_2Q) + \mu_2 O_1O_2$$

or $$\mu_1 \left[(PO^2 + AO^2)^{\frac{1}{2}} + (OQ^2 + AO^2)^{\frac{1}{2}} \right]$$
$$= \mu_1 (PO + OQ) + (\mu_2 - \mu_1)(t_1 + t_2).$$

or $$\mu_1 \left[PO\left(1 + \frac{AO^2}{2PO^2} + \ldots\right) + OQ\left(1 + \frac{AQ^2}{2OQ^2} + \ldots\right) \right]$$
$$= \mu_1 (PO + OQ) + (\mu_2 - \mu_1)(t_1 + t_2).$$

or $$\mu_1 PO + \mu_1 \frac{AO^2}{2PO} + \mu_1 OQ + \mu_1 \frac{AO^2}{2OQ}$$
$$= \mu_1 PO + \mu_1 OQ + (\mu_2 - \mu_1)(t_1 + t_2)$$

Putting $AO = y$, $PO = u$, $OQ = v$, r_1 = radius of curvature of first surface and r_2 = that of second surface, we have

$$\frac{\mu_1 y^2}{2u} + \frac{\mu_1 y^2}{2v} = (\mu_2 - \mu_1)\left(\frac{y^2}{2r_1} + \frac{y^2}{2r_2}\right),$$

Since $t_1 = \frac{y^2}{2r_1}$ and $t_2 = \frac{y^2}{2r_2}$, (*cf*, Sagitta theorem)

$$\therefore \quad \frac{\mu_1}{v} - \frac{\mu_1}{u} = (\mu_2 - \mu_1)\left(\frac{1}{r_1} + \frac{1}{r_2}\right)$$

Putting signs according to our conventions, we have

$$\frac{\mu_1}{v} - \frac{\mu_1}{u} = (\mu_2 - \mu_1)\left(\frac{1}{r_1} - \frac{1}{r_2}\right),$$

or
$$\frac{1}{v} - \frac{1}{u} = \left(\frac{\mu_2 - \mu_1}{\mu_1}\right)\left(\frac{1}{r_1} - \frac{1}{r_2}\right),$$

from which the equations

$$\frac{1}{f} = \left(\frac{\mu_2 - \mu_1}{\mu_1}\right)\left(\frac{1}{r_1} - \frac{1}{r_2}\right)$$

and
$$\frac{1}{v} - \frac{1}{u} = \frac{1}{f}$$ follow at once.

Example: *Convex lens of crown glass, refractive index 1.50 is placed in (a) water, $\mu = 1.33$ and (b) carbon bisulphide, $\mu = 1.65$. Determine in each case whether the lens behaves as a converging lens or diverging lens and find its focal length. Assume the focal length of the lens in air is 15 cm.*

Let f, f_w, f_c be the focal lengths of the lens in air, in water and in carbon bisulphide respectively, then

$$\frac{1}{f} = (\mu - 1)\left(\frac{1}{r_1} - \frac{1}{r_2}\right),$$

or
$$\frac{1}{15} = (1.50 - 1)\left(\frac{1}{r_1} - \frac{1}{r_2}\right) = 0.5\left(\frac{1}{r_1} - \frac{1}{r_2}\right) \quad .. (i)$$

Also $$\frac{1}{f_w} = \left(\frac{\mu_2 - \mu_1}{\mu_1}\right)\left(\frac{1}{r_1} - \frac{1}{r_2}\right),$$

[where $\mu_2 = 1.50$ and $\mu_1 = 1.33$]

whence, $$\frac{1}{f_w} = \left(\frac{1.50 - 1.33}{1.33}\right)\left(\frac{1}{r_1} - \frac{1}{r_2}\right)$$

$$= \frac{0.17}{1.33}\left(\frac{1}{r_1} - \frac{1}{r_2}\right) \quad \text{... (ii)}$$

and $$\frac{1}{f_c} = \left(\frac{\mu_2 - \mu_3}{\mu_3}\right)\left(\frac{1}{r_1} - \frac{1}{r_2}\right)$$

[where $\mu_2 = 1.50$ and $\mu_3 = 1.65$]

$\therefore$ $$\frac{1}{f_c} = \left(\frac{1.50 - 1.65}{1.65}\right)\left(\frac{1}{r_1} - \frac{1}{r_2}\right)$$

$$= \frac{0.15}{1.65}\left(\frac{1}{r_1} - \frac{1}{r_2}\right) \quad \text{... (iii)}$$

Dividing (*i*) by (*ii*), we get $\frac{f_w}{15} = +0.5 \times \frac{1.33}{0.17}$

or $$f_w = +\frac{0.5 \times 1.33 \times 15}{0.17} = +58.7 \text{ cm.}$$

The negative sign indicates that the lens behaves as a diverging lens.

Dividing (*i*) by (*iii*)

$$\frac{f_e}{15} = -\frac{0.5 \times 1.65}{0.15},$$

$\therefore$ $$f_e = -\frac{0.5 \times 1.65 \times 15}{0.15} = -82.5 \text{ cm.}$$

The negative sign indicates that the lens behaves as a diverging lens.

Example: *A biconvex lens of radii of curvature 30 0 cm each is placed on the surface of water so that the lower surface is only immersed. An illuminated object 40 cm deep is observed vertically through the lens and water. Find the position of the image. Refractive index of air glass = 3/2 and for air-water = 4/3.*

When there are media of refractive indices μ_1 and μ_2 on either side of the lens of refractive index μ_2 then tor refraction at the first surface

$$\frac{\mu_2}{v'} - \frac{\mu_1}{u} = \frac{\mu_2 - \mu_1}{r_2} \quad \text{... } (a)$$

and for refraction at the second surface

$$\frac{\mu_2}{v} - \frac{\mu_2}{v'} = \frac{\mu_3 - \mu_2}{r_2} \quad \text{... } (b)$$

neglecting the thickness of the lens.

Adding (*a*) and (*b*)

$$\frac{\mu_3}{v} - \frac{\mu_2}{u} = \frac{\mu_3 - \mu_1}{r_1} + \frac{\mu_3 - \mu_2}{r_2} \quad \text{... } (c)$$

Here, $\mu_1 = \frac{4}{3},\ \mu_2 = \frac{3}{2},\ \mu_3 = 1.0$

Conventionally $u = -40$ cm, $r_1 = 30$ cm and $r_2 = -30$ cm.

Substituting the values in equation (*c*), we have

$$\frac{1.0}{v} - \frac{\frac{4}{3}}{-40} = \frac{\frac{3}{2} - \frac{4}{3}}{30} + \frac{1.0 - \frac{3}{2}}{-30}$$

or

$$\frac{1}{v} + \frac{1}{30} = \frac{1}{180} + \frac{1}{60},$$

or
$$\frac{1}{v} = \frac{1}{180} + \frac{1}{60} - \frac{1}{30}$$
$$= \frac{1+3-6}{180} = -\frac{2}{180},$$

whence $v = -90$ cm

That is, the image will be on the same side as the object at a distance of 90 cm from the lens and is virtual.

Example: *A convex lens of focal length 20 cm and made of glass ($\mu = 1.50$) is immersed in water ($\mu = 1.33$). Find the change in the focal length of the lens.*

For lens in air,

$$\frac{1}{f} = (\mu - 1)\left(\frac{1}{r_1} - \frac{1}{r_2}\right),$$

where $\mu = 1.50$ and $f = 20$ cm.

$$\therefore \quad \frac{1}{20} = (1.50 - 1)\left(\frac{1}{r_1} - \frac{1}{r_2}\right) = 0.5\left(\frac{1}{r_1} - \frac{1}{r_2}\right), \quad \text{... } (i)$$

For lens in water,

$$\frac{1}{f_1} = \left(\frac{1.50 - 1.33}{1.33}\right)\left(\frac{1}{r_1} - \frac{1}{r_2}\right)$$
$$= \frac{0.17}{1.33}\left(\frac{1}{r_1} - \frac{1}{r_2}\right) \quad \text{... } (iii)$$

Dividing (*i*) by (*ii*)

$$\frac{f_1}{20} = \frac{0.5 \times 1.33}{0.17},$$

or
$$f_1 = \frac{0.5 \times 1.33 \times 20}{0.17} = +\ 78.2 \text{ cm.}$$

The positive sign indicates that the lens behaves as a converging lens of focal length 78.2 cm so that the increase in focal length is equal to 78.2 – 20 = 58.2 cm.

Example: *A plano-convex lens of focal length 15 cm is fixed at the centre of one side of a box with convex surface outside. An object 5 cm high is placed at 25 cm from the lens. Find the position and size of the image, when the box is (a) empty, (b) filled with water.*

$$^a\mu_g = \frac{3}{2};\ ^a\mu_w = \frac{4}{3}.$$

(*a*) When the box is empty,

$$\frac{1}{v} - \frac{1}{u} = \frac{1}{f},$$

or $$\frac{1}{v} - \frac{1}{-25} = \frac{1}{15}, \qquad \text{whence } v = 37.5 \text{ cm.}$$

Now, $$\frac{\text{size of image}}{\text{size of object}} = \frac{v}{u} = \frac{37.5}{25}$$

whence, size of image $= \dfrac{37.5 \times 5}{25} = 7.5$ cm.

Since the lens is plano-convex with convex surface outside

$$\therefore \qquad \frac{1}{f} = (\mu - 1)\left(\frac{1}{r_1} - \frac{1}{\infty}\right)$$

or $$\frac{1}{15} = \frac{0.5}{r_1}, \text{ whence } r_1 = \frac{15}{2} \text{ cm:}$$

(*b*) When there is water on the other side, with usual notation, we have

$$\frac{\mu_3}{v} - \frac{\mu_1}{u} = \frac{\mu_2 - \mu_1}{r_1} - \frac{\mu_3 - \mu_2}{r_2}$$

Substituting the values in this equation, we get

$$\frac{\frac{4}{3}}{v} - \frac{1}{-25} = \frac{\frac{3}{2} - 1}{\frac{15}{2}} - \frac{\frac{4}{3} - \frac{3}{2}}{\infty} = \frac{1}{15},$$

or
$$\frac{4}{3v} = \frac{1}{15} - \frac{1}{25} = \frac{2}{75},$$

or
$$v = 50 \text{ cm}$$

$$\text{Size of image} = \frac{50}{25} \times 5 = 10 \text{ cm.}$$

Techniques of Deviation

Derivation of Lens Formula by Deviation Method: This method is based on the prismatic properties of a lens, which may be regarded as a series of prisms of small angles which increase in size of the refracting angles from the centre of the lens outwards. In the case of a convex lens these refracting angles are pointing away from the axis and in the case of a concave lens, they are pointing inwards. We know that for a prism of small angle a, the deviation *S*, is give by

$$\delta = (\mu - 1)\ \alpha,$$

where μ is the refractive index of the material of the prism.

In deriving the lens formula we shall make use of the paraxial rays and assume that the medium on both sides of the lens is the same unless otherwise stated.

If a ray *PA* parallel to the axis is incident at a point *A* distant *h* from the centre of the lens, on emerging it meets the axis in *F* and the deviation produced is given by

$$\delta = \frac{h}{f},$$

where f is the focal length.

Next, let *P* be a point object lying on the principal axis. A ray *PA* incident at the same distance *h* from the centre of the lens suffers the deviation δ_1 and after refraction cuts the axis in *Q*. Since the deviation is independent of the angle of incidence- for paraxial rays, $\delta_1 = \delta$.

From the geometry of,

$$\delta = \frac{h}{f},$$

and $$\delta_1 = \alpha + \beta = \frac{h}{-u} + \frac{h}{v}$$

$\therefore$ $$\frac{h}{f} = \frac{h}{-u} + \frac{h}{v}$$

or $$\frac{1}{f} = \frac{1}{v} - \frac{1}{u}.$$

The fundamental lens formula can be derived as under :

We have seen above that the angle of deviation is given by

$$\delta = \frac{h}{f},$$

where h is the height of the point of incidence above the axis.

Now, $$\delta = (\mu - 1)\,\alpha,$$

where α is the refracting angle of the prism forming the corresponding portion of the lens α, therefore, is the angle between the tangent planes to the lens faces at the distance h from the axis.

We find that a is also the angle between the radial drawn to the lens faces at the same distance.

so that $$\alpha = \theta + \phi,$$

whence $$\delta = (\mu - 1)\,(\theta + \phi).$$

But $$\delta = \frac{h}{f}, \text{ therefore } \frac{h}{f} = (\mu - 1)\,(\theta + \phi)$$

Now, conventionally $\theta = \dfrac{h}{r_1}$ and $\phi = -\dfrac{h}{r_2}$.

$$\therefore \qquad \frac{h}{f} = (\mu - 1)\left(\frac{h}{r_1} - \frac{h}{r_2}\right),$$

or
$$\frac{1}{f} = (\mu - 1)\left(\frac{1}{r_1} - \frac{1}{r_2}\right).$$

Example: *A thin air double concave lens bounded by very thin glass walls with radii of curvature 24 cm for each surface separates a long tank into two compartments one of which is filled with water of refractive index 1.33 and the other with a liquid of refractive index 1.67. If a parallel beam of light traversing the water strikes the lens, find where it is focused ?*

For refraction at the first surface,

$$\frac{\mu_2}{v'} - \frac{\mu_1}{u} = \frac{\mu_2 - \mu_1}{r_1}.$$

For refraction at second surface,

$$\frac{\mu_2}{v} - \frac{\mu_2}{v'} = \frac{\mu_3 - \mu_2}{r_2}.$$

Adding
$$\frac{\mu_3}{v} - \frac{\mu_1}{u} = \frac{\mu_2 - \mu_1}{r_1} + \frac{\mu_3 - \mu_2}{r_2}.$$

Since incident light is a parallel beam, $u = -\infty$, $v = f$,

$$\therefore \qquad \frac{\mu_3}{f} = \frac{\mu_2 - \mu_1}{r_1} + \frac{\mu_3 - \mu_2}{r_2}.$$

Here, $\mu_3 = 1.67$; $\mu_2 = 1.00$; $\mu_1 = 1.33$; $r_1 = -24$; $r_2 = +24$.

$$\therefore \qquad \frac{1.67}{f} = \frac{1.00 - 1.33}{-24} + \frac{1.67 - 1.00}{24}$$

$$= \frac{0.33}{24} + \frac{0.67}{24} = \frac{1}{24}$$

whence $f = 1.67 \times 24 = 40.1$ cm.

Hence the image is 40.1 cm from the lens in the second liquid.

Minimum Distance between the Object and its Real Image in a Convex Lens: In order to obtain the real image of an object on the screen by means of a convex lens the object and the screen must be placed at a distance apart which is greater than a certain minimum value. Let x be this distance. Then,

$$x = OP + OQ = -u + v \quad \text{...}(i)$$

and we have to determine the minimum value of x subject to the condition that

$$\frac{1}{v} - \frac{1}{u} = \frac{1}{f} \quad \text{...}(ii)$$

Eliminating u from these equations, we have

$$\frac{1}{v} - \frac{1}{v-x} = \frac{1}{f}$$

or $$\frac{v-x-v}{v(v-x)} = \frac{1}{f}$$

or $$v(v-x) = -xf$$

or $$v^2 - vx + xf = 0$$

Solving the quadratic for v, we get,

$$v = \frac{x \pm \sqrt{x^2 - 4xf}}{2} \quad \text{...}(iii)$$

For v to be real

$$x^2 - 4xf = + \text{ve or zero.}$$

$\therefore$ $$x^2 \geqq 4xf$$

or $$x \geqq 4f.$$

That is, the *minimum* value of x is $4f$.

or *alternatively*, we may use the method of differential calculous.

Again $$x = v - u$$

and $$\frac{1}{v} - \frac{1}{u} = \frac{1}{f}.$$

Eliminating v from these equations, we have

$$x = \frac{uf}{u+f} - u.$$

Differentiating, $$\frac{dx}{du} = \frac{(u+f)f - uf}{(u+f)^2} - 1 = \frac{f^2 - (u+f)^2}{(u+f)^2}$$

$$= \frac{-u(u+2f)}{(u+f)^2}.$$

Now for x to be a minimum or a maximum

$$\frac{dx}{du} = 0,$$

so that, $$u = 0,$$

or $$u + 2f = 0.$$

The value $u = 0$ is of no physical consequence, as it means that the object is in contact with the lens. The second value $u = -2f$, when substituted in

$$\frac{1}{v} - \frac{1}{u} = \frac{1}{f}, \text{ gives } v = 2f.$$

Then, $$x = 2f - (-2f) = 4f.$$

Differentiating again and simplifying, we get

$$\frac{d^2x}{du^2} = \frac{-2f^2}{(u+f)^2}$$

Substituting $u = 2f$, gives

$$\frac{d^2x}{du^2} = \frac{2}{f}, \text{ a positive value, since } f \text{ is + ve.}$$

Therefore, the value of $x = 4f$ is *minimum* under these conditions.

Note: From the relation (*iii*) above, it is obvious that there are two values of v given by

$$v_1 = \frac{x + \sqrt{x^2 - 4xf}}{2}$$

and
$$v_2 = \frac{x - \sqrt{x^2 - 4xf}}{2}$$

Since, $u = v - x$, we have

$$u_1 = v_1 - x = \frac{-x + \sqrt{x^2 - 4xf}}{2}$$

and
$$u_2 = v_2 - x = \frac{-x - \sqrt{x^2 - 4xf}}{2}$$

Thus, we have

$$u_1 = -v_2 \text{ and } u_2 = -v_1.$$

Hence, in general there are two coaxial positions of a convergent lens which will produce on a fixed screen, a sharp image of a fixed object.

Two Thin Lenses in Contact: Sometimes for removing certain aberrations we have to make use of two lenses placed in contact. We shall therefore, find the combined focal length of such a lens system. Let two thin lenses of focal lengths f_1 and f_2, be placed in contact so that both have a common axis. An object at a distance u from the first lens of focal length f_1, gives an image at a distance v' from it, so that

$$\frac{1}{v'} - \frac{1}{u} = \frac{1}{f_1} \quad \text{... } (a)$$

This image serves as a virtual object for the second lens and since the lenses are thin and in contact the final image is formed at a distance v from the lens system given by

$$\frac{1}{v} - \frac{1}{v'} = \frac{1}{f_2} \qquad \text{... (b)}$$

Adding (a) and (*b*), v' is eliminated and we get

$$\frac{1}{v} - \frac{1}{u} = \frac{1}{f_1} + \frac{1}{f_2}.$$

But $$\frac{1}{v} - \frac{1}{u} = \frac{1}{F}$$

where F is the focal length of the combination.

$$\frac{1}{F} = \frac{1}{f_1} + \frac{1}{f_2}. \qquad \text{... (14)}$$

In other words, the reciprocal of the focal length of a thin lens combination is equal to the sum of the reciprocals of the focal lengths of the component lenses.

The single lens of focal length F, when placed in the position occupied by the lens combination produces an image of a given object in the same position and of the same size as that produced by the combination.

Since power of a lens, $P = 1/f$, we obtain for the power of the combination

$$P = P_1 + P_2. \qquad \text{... (15)}$$

So that when two thin lenses are placed in contact, *the power of the combination is equal to the sum of the powers of the component lenses.*

In general, if there are n thin lenses, we have

$$\frac{1}{F} = \frac{1}{f_1} + \frac{1}{f_2} + \frac{1}{f_3} + \dots\dots\dots + \frac{1}{f_n} \qquad \text{... (16)}$$

and $$P = P_1 + P_2 + ... + P_n. \quad ... (17)$$

The relations (14), (15), (16) and (17) hold good whether, the lenses are all convex or all concave or some convex and some concave.

The Refraction

A lens is a portion of a transparent medium bounded by two spherical surfaces or one plane and the other spherical. The line which passes through the centres of curvature of the two faces or which passes through the centre of curvature of one face and is perpendicular to the other face is called the principal axis of the lens.

The points at which the axis cuts the faces of the lens are the vertices and the distance between them is the thickness of the lens. In this chapter we shall deal with this lenses, *i.e., lenses* of which thickness can be neglected in comparison with object and image distances. Although there are various types of lenses, yet they can be classed into two main groups according as their thickness at the centre is greater or less than at the edges. The former are called converging lenses and the latter diverging lenses.

In considering refraction through a lens we shall confine our attention to rays which make small angles with the axis and adopt the approximations and conventions of sign.

Refraction through a Thin Lens: Consider a thin convex lens L of refractive index μ placed in air. Let an object P, lie on the principal axis to the left of the lens. If Q' be the position of the image formed by refraction at the first surface, then Q' serves as the virtual object for refraction at the second surface giving rise to the final image at Q.

If O, O' are the vertices, for a thin lens, its thickness $(t) = OO'$ is neglected. Let $OP = u$, $OQ' = v'$ and $O'Q = v$ and r_1, r_2 the radii of curvature of the first and second surface respectively. Then, assuming that a lens of refractive index μ is placed in air. Then for refraction at the first surface, we have

$$\frac{\mu}{v'} - \frac{1}{u} = \frac{\mu - 1}{r_1} \quad \text{... }(a)$$

and for refraction at the second surface measuring the distance of image from O' and keeping in view that rays are now going from a denser to a rarer medium,

$$\frac{\frac{1}{\mu}}{v} - \frac{1}{v' - t} = \frac{\frac{1}{\mu} - 1}{r_2}.$$

Neglecting t, since the lens is thin and simplifying, we have

$$\frac{1}{v} - \frac{\mu}{v'} = \frac{1 - \mu}{r_2} \quad \text{... }(b)$$

Adding (*a*) and (*b*), we get

$$\frac{1}{v} - \frac{1}{u} = (\mu - 1)\left(\frac{1}{r_1} - \frac{1}{r_2}\right) \quad \text{... (1)}$$

The points *P* and *Q* thus connected, are known as conjugate foci and equation (1) is known as thin lens formula. *Or more generally,* if refractive index of the lens is μ_2 that of the medium on the left and right, μ_1 and μ_2 respectively we have, for refraction at the first and the second surfaces respectively

$$\frac{\mu_2}{v'} - \frac{\mu_1}{u} = \frac{\mu_2 - \mu_1}{r_1} \quad \text{... }(c)$$

and $$\frac{\mu_2}{v} - \frac{\mu_2}{v'} = \frac{\mu_2 - \mu_2}{r_2} \quad \text{... }(d)$$

Adding (*c*) and (*d*), we get

$$\frac{\mu_2}{v} - \frac{\mu_1}{u} = \frac{\mu_2 - \mu_1}{r_1} + \frac{\mu_2 - \mu_2}{r_2} \quad \text{... (1}a\text{)}$$

Assuming the lens to be placed in a uniform medium, so that

$$\mu_1 = \mu_2,$$

then
$$\frac{\mu_1}{v} - \frac{\mu_1}{u} = \frac{\mu_2 - \mu_1}{r_1} + \frac{\mu_1 - \mu_2}{r_2}$$

$$= (\mu_2 - \mu_1)\left(\frac{1}{r_1} - \frac{1}{r_2}\right).$$

or
$$\frac{1}{v} - \frac{1}{u} = \left(\frac{\mu_2 - \mu_1}{\mu_1}\right)\left(\frac{1}{r_1} - \frac{1}{r_2}\right). \quad \text{... (1b)}$$

The reader is expected to derive the relations 1, (1*a*) and (1*b*) for a concave lense himself.

First and Second Principal Foci of a Thin Lens: The limiting positions of the object and the image formed by refraction an infinite distance are called the an infinite distance are called the first and second principal foci of lens. These points are denoted by F_1 and F_2 and their distances from the centre of the lens are known as the first and the second local lengths of the lens. The first and the second principal foci of a convex and a concave lens respectively. The foci are *real* for a convex lens and *virtual* for a concave lens. Denoting their distances from the centre of the lens by f_1 and f_2, we have

$u = f_1$, when $v = \infty$ and $v = f_2$, when $u = \infty$.

Putting these values in equation (1), we have

$$\frac{1}{\infty} - \frac{1}{f_1} = (\mu - 1)\left(\frac{1}{r_1} - \frac{1}{r_2}\right)$$

or
$$-\frac{1}{f_1} = (\mu - 1)\left(\frac{1}{r_1} - \frac{1}{r_2}\right) \quad \text{... (2)}$$

and
$$\frac{1}{f_2} - \frac{1}{\infty} = (\mu - 1)\left(\frac{1}{r_1} - \frac{1}{r_2}\right)$$

or
$$\frac{1}{f_2} = (\mu - 1)\left(\frac{1}{r_1} - \frac{1}{r_2}\right) \qquad \text{... (2a)}$$

From these equations it is obvious that the first and second focal lengths of a lens are *numerically* equal when the lens is placed in a uniform medium. In the subsequent sections, when we speak of the focal length of a lens, we shall mean the second local length when the lens is placed in air and denote it by *f*. Hence, equation (1) may be written as

$$\frac{1}{v} - \frac{1}{u} = \frac{1}{f} = (\mu - 1)\left(\frac{1}{r_1} - \frac{1}{r_2}\right) \qquad \text{... (3)}$$

Equation (3) gives a fundamental thin lens relation expressing the focal length in terms of the radii of curvature and the refractive index of the material of the lens and is often known as Lens makers' formula.

Since $\mu > 1$, the sign of focal length depends upon that of

$$\left(\frac{1}{r_1} - \frac{1}{r_2}\right).$$

If a lens of refractive index μ_2 is placed in a uniform medium of refractive index μ_1, the focal length is given by

$$\frac{1}{f} = \left(\frac{\mu_2 - \mu_1}{\mu_1}\right)\left(\frac{1}{r_1} - \frac{1}{r_2}\right) \qquad \text{... (3a)}$$

and as a consequence of equations (2) and (2*a*) for the first and the second principal focal lengths respectively, the above equation may be written as

$$\frac{f_2}{v} + \frac{f_1}{u} = 1. \qquad \text{... (4)}$$

If, however, a lens of refractive index μ_2 is placed in a medium of refractive index μ_1 on the left and that of refractive index μ_3 on the right, then keeping in mind the definitions of the first and second focal lengths we can easily deduce from equation (1*a*) that

$$\frac{\mu_3}{v} - \frac{\mu_1}{u} = -\frac{\mu_1}{f_1} = \frac{\mu_3}{f_2} \qquad \text{... (5)}$$

and

$$\frac{f_1}{f_2} = -\frac{\mu_1}{\mu_2} \qquad \text{... (5a)}$$

That is, the first *and the second focal lengths are in the ratio of the refractive indices of the media in which the corresponding focal points lie.*

Lens with Strength

The expression

$$(\mu - 1)\left(\frac{1}{r_1} - \frac{1}{r_2}\right),$$

which by equation (3) is equal to the reciprocal of the focal length, is called the power of the lens. Denoting it by P, we have

$$P = \frac{\mu - 1}{r_1} + \frac{1 - \mu}{r_2}, \qquad \text{... (6)}$$

from which it follows that the *power of a lens is the sum of the powers of its refracting surfaces.*

According to the conventions of sign used in the text, for a biconvex lens in air, r_1 is positive and r_2 is negative, whence the focal length and power of a convex lens are positive. For a biconcave lens in air, r_1 is negative and r_2 is positive, so the focal length and power are both negative.

In practical optics a lens is of *unit* power if it brings parallel rays to focus at a distance of 1 metre. This unit of converging or diverging power is called the dioptre. Thus a converging lens of 1 metre focal length is said to have a power +1 D and that of 25 cm focal length + 4D. Similarly, a concave lens of 25 cm focal length has a power – $4D$ and so on.

Optical Centre of a Lens: Let C_1 and C_2 be the centres of curvature of the faces of a biconvex lens placed in a uniform medium so that the principal axis cuts the leas surfaces in O_1 and O_2. Draw two parallel radii C_1Q and C_2R intersecting the leas surfaces in Q and R respectively, then the tangents at Q and R will obviously *be* parallel to each other. Hence, a ray PQ which on refraction traverses the path QR will emerge parallel to PQ. The point O, where the ray QR cuts the axis, is called the optical centre of the lens. To determine the position of O consider the similar Δs C_1PQ and C_2OR

Here, $$\frac{OC_1}{OC_2} = \frac{C_1Q}{C_2R} = \frac{C_1C_2}{O_2C_2} = \frac{r_1}{r_2}.$$

Also since $$\frac{O_1C_1}{O_2C_2} = \frac{OC_1}{OC_2} = \frac{O_1C_1 - OC_1}{O_2C_2 - OC_2} = \frac{O_10}{O_20}.$$

$$\therefore \quad \frac{O_10}{O_20} = \frac{r_1}{r_2}. \qquad \text{... (7)}$$

that is, for a given lens the position of O is fixed. Hence, any ray which emerges from the lens in a direction parallel to the incident ray, passes through the point O or *vice versa*. If the lens is thin, we can almost say that a ray passing through the optical centre goes straight. This property is used in the construction of the images of objects.

Construction of the Image of a Small Object by a Lens: Let PQ, be a small object placed perpendicular to the axis of a lens. A ray PA incident parallel to the axis, after refraction proceeds in the direction AF_2 or F_2A and a ray PB incident in the direction PF_1, after refraction becomes parallel to the axis. Let these two rays intersect or appear to do so, then P' is the image of P and $P'Q'$ perpendicular to the axis is the image of PQ. Since a ray passing through the optical centre O of the lens suffers no deviation, P and P' will also lie on the line passing through

(a) *Transverse or Lateral Magnification:* It is defined as the ratio of any transverse dimension of the image to the corresponding transverse dimension of the object. Denoting it by m, we have

$$m = \frac{h_2}{h_1} = -\frac{P'Q'}{PQ} = -\frac{BO}{PQ} = -\frac{F_1O}{F_1Q}$$

$$= -\frac{F_1O}{OQ - OF_1} = -\frac{+f}{-u-f} = \frac{f}{u+f} \quad \text{... (8)}$$

Also,

$$\mu = \frac{P'Q'}{AO} = -\frac{F_2Q'}{F_2O} = -\frac{OQ' - OF_2}{F_2O}$$

$$= \frac{v-f}{f} = \frac{f-v}{f} \quad \text{... (9)}$$

Thus, transverse magnification varies *inversely* as the distance of the object or *directly* as the distance of the image from the corresponding focus.

Another expression for the transverse magnification is

$$m = \frac{h_2}{h_1} = \frac{P'Q'}{PQ} = -\frac{OQ'}{OQ} = \frac{v}{-u} = \frac{v}{u} \quad \text{... (10)}$$

This result is also obtained by considering the magnification at each surface. If m_1 and m_2 be their respective magnifications, then

$$m_1 = \frac{v'}{u} \div \frac{u}{1} = \frac{v'}{\mu u}$$

and

$$m_2 = \frac{v}{1} \div \frac{v'-t}{\mu} = \frac{\mu v}{v},$$ when t is negligible.

Hence the resultant magnification

$$m = m_1 \times m_2 = \frac{v'}{\mu u} \times \frac{\mu v}{v'} = \frac{v}{u}.$$

(b) *Longitudinal or Axial Magnification:* It is defined as the ratio of the depth of the image to the corresponding depth of the object. Since all objects are essentially three dimensional, it is desirable to study the effect of longitudinal magnification. Denoting it so that differentiating this relation, we have

$$m' = \frac{\delta v}{\delta u},$$

where $$\frac{1}{v} - \frac{1}{u} = \frac{1}{f},$$

so that differentiating this relation, we have

$$-\frac{\delta v}{v^2} - \left(-\frac{\delta u}{u^2}\right) = 0,$$

or $$\frac{\delta v}{v^2} - \frac{\delta u}{u^2} = 0,$$

and therefore $$m' = \frac{\delta v}{\delta u} = \frac{v^2}{u^2} = \left(\frac{v}{u}\right)^2 = m^2 \qquad \text{...(11)}$$

(c) *Angular Magnification:* It is defined as the ratio $\frac{\tan \alpha_2}{\tan \alpha_1}$ where a_1 and a_2 are the slope angles of the object ray and the image ray respectively. It is denoted by g.

$$\therefore \qquad \gamma = \frac{\tan \alpha_2}{\tan \alpha_1}$$

By Lagrange-Helmholtz equation,

$$\mu_1 h_1 \sin \alpha_1 = \mu_2 h_2 \sin \alpha_2$$

or $$\mu_1 h_1 \tan \alpha_1 = \mu_2 h_2 \tan \alpha_2,$$

[$\because$ α_1 and α_2 are small]

Since $$\mu_1 = \mu_2$$

$$h_1 \tan \alpha_1 = h_2 \tan \alpha_2.$$

$\therefore$ angular magnification

$$\gamma = \frac{\tan \alpha_2}{\tan \alpha_1} = \frac{h_1}{h_2} = \frac{1}{m}, \qquad \text{... (12)}$$

that is, *the angular magnification is reciprocal of the lateral magnification produced by a lens.*

From relations (11) and (12), we have

$$m' \times \gamma = m^2 \times \frac{1}{m} = m,$$

that is, longitudinal magnification × angular magnification = transverse magnification.

Polarisation and Polarimeter

Polarimeter of Laurent

The essential parts of a Laurent's polarimeter. It consists of two nicol prisms *P* and *A* mounted in brass tubes placed some distance apart and capable of rotation about a common axis, a half shade device *H*, so ingeniously invented by Laurent which divides the field of the polarised light emerging out of the nicol *P* into two parts generally of unequal brightness. A glass tube *D* with ground ends, against which fit plane glass plates, contains the active solution, say, that of sugar. It is mounted between the two nicols. All the tubes are arranged in one line and are mounted on a rigid iron base.

Monochromatic light from a source S, rendered slightly convergent by a convex lens *L*, is made to fall upon the polarising nicol *P*. After passing through *P* the light is rendered plane-polarised with its vibrations in the principal plane of the nicol. This plane-polarised light before passing through the analysing nicol *A* is made to traverse the half-shade device *H* and the tube *D* containing the active solution.

The emergent light is viewed through a Galilean telescope *T*. The analysing nicol *A* can be rotated about its axis and its rotation can be accurately measured on a circular scale graduated in degrees with the help of vernier (V.S.) attached to it.

In some polarimeters, the vernier scale is fixed and the circular graduated scale is capable of rotation with the nicol.

Half-shade Device: It was to overcome the inability of the eye to judge the exact position of extinction, when the two nicols are adjusted in the crossed position that half-shade device had its origin. It consists of two semi-circular plates, one made of glass and the other of quartz. Both glass and quartz halves are cemented together. The quartz plate is cut parallel to its optic axis, which is parallel to *YY′*, the line joining the two plates. It is a half-wave plate for sodium light, *i.e.*, its thickness is such that it introduces a path difference of half a wavelength between the ordinary (*O*) and extraordinary (*E*) waves in sodium light. In other words, in passing through the plate, ordinary wave *gains* half a wavelength on the extraordinary wave. The glass plate should be of such a thickness that it transmits the same amount of light as quartz plate does.

To understand the action of Laurent's plate let us suppose that the principal plane of the polarising nicol makes an angle θ with the optic axis of the quartz plate, *i.e.*, the principal plane of the polarising nicol is parallel to PP_1. When the ordinary monochromatic light after passing through the polarising nicol is incident normally on the half-shade plate the direction of vibrations of the plane-polarised light is parallel to PP_1. On passing through the glass half, the vibrations will remain along the same direction, *i.e.*, PP_1, because glass is not doubly refracting; but in the quartz-half, the vibrations will be split up into *E* and *O* components polarised at right angles to each other.

The B component is parallel to the optic axis of the quartz *(XT′)* while the *O* component is perpendicular to it, *i.e.*, along *XX′*. These two rectangular components travel with unequal speeds within quartz and when they arrive at the opposite face, they differ in phase by π. Because of this phase difference of x the direction of the O component in quartz is reversed. Thus the two components OA and OC of the original incident vibrations along OP, on passing through quartz become OA and OD, the latter being the reverse of OC. These two component vibrations on emergence from quartz combine to form a linear resultant vibration, whose direction will be along OQ inclined with the optic axis OA (Y-axis) at the same angle θ as the incident vibration along OP with that axis but on the other side of it,

i.e., $\angle AOQ = \angle AOP$.

Thus the effect of the quartz plate has been to rotate the plane of vibration by an angle 2θ.

Now if the principal plane of the A-nicol is parallel to P_1OP the plane-polarised light through the glass-half will pass unobstructed while that through the quartz-half is partially obstructed so that the glass-half will appear brighter than the quartz-half. On the other band, if the principal plane of the A-nicol is parallel *to* QQ_1 the quartz-half will appear brighter than the glass-half for the same reason as above.

It is this property of the half-shade device of dividing the field into two halves of unequal intensity that has made this polarimeter so successful.

When the principal plane of the A-nicol is parallel to OA (Y-axis)—a line which bisects the $\angle POQ$ – the emergent vibrations from both the glass-half and quartz-half are equally inclined to the principal plane of the A-nicol and hence the two components make the two halves *equally bright*.

When the principal plane of the A-nicol is parallel to OC (X - axis)—at right angles to the optic axis of the quartz – again the two halves are *equally bright* since OP and OQ are equally inclined to X-axis and the transmitted amplitudes OC and OD are equal. But the intensity in both parts of the field in this latter case being $I = I_0 \sin^2 \theta$, is small as compared with I_0 if $\theta \to 0$. This is the position of equal darkness in the two portions of the field. The variation in the intensity of the field as the *A*-nicol is rotated through a small angle about this position. A slight rotation of the -*A*-nicol from the setting say, in the clockwise direction causes a rapid increase in the transmitted amplitude *OD* from the quartz-half and a decrease in the transmitted amplitude *OC* from the glass-half. Thus, the quartz part appears brighter than the glass part. On the other hand, a slight rotation of the *A*-nicol in the anti-clockwise direction causes the glass part to appear brighter than the quartz part. It is in this setting of the analysing nicol that the eye can easily detect a slight variation in the intensity of light and the readings are therefore taken for this position. The advantage of this device over the extinction method lies in the fact that *the slightest rotation of the A-nicol from this critical position produces a considerable variation in the brightness of the two halves.*

The light coming out of the *A*-nicol is examined by a telescope. The position of the *A-nicol* is adjusted by rotating the tube containing, the *A*-nicol and the amount of rotation is measured with the help of the attached vernier scale.

The angle 2θ between *OP* and *OQ* is known as half-shadow angle and can be varied by rotating the *P*-nicol about its own axis. The sensitiveness of the apparatus depends upon the magnitude of this angle and there is certain *minimum* half-shadow angle for which this polarimeter is very sensitive.

It may be noted that this polarimeter is suitable for only one particular wavelength for which this half-shade device is made.

Example: *A 20 cm long tube containing sugar solution is placed between crossed nicols and is illuminated by light of wavelengh 6 × 10^{-5} cm. If the specific rotation is 60° and optical rotation produced is 12°, what is the strength of the solution ?*

The specific rotation *S* at a given temperature *t* and far a given wavelength of light λ used is given by

$$S_t\lambda = \frac{\theta}{I \times d},$$

where θ is the rotation, *d* is the density of the solution in g/cm³ and *I* is its length in decimetres.

Hence, $S = 60°$; $I = 2$ decimetres; $\theta = 12°$

and $\lambda = 6 \times 10^{-5}$ cm

$$\therefore \qquad \delta = \frac{\theta}{I \times S} = \frac{12}{2 \times 60} = \frac{1}{10}.$$

∴ it is a 10 per cent solution of sugar, *i.e.*, 1 g of sugar is dissolved in 10 cm³ of solution.

Polarimeter of Lippich

Laurent's half-shade device is suitable only for one particular wavelength for which the quartz half-wave plate is made. To overcome this difficulty other methods have been employed. Out of these *Lippich polarimeter* is one which needs a special mention.

In this case the polarise which consists of two nicols P_1 and P_2 with their principal planes inclined to each other at a small angle θ is placed in front of the aperture S_1. The principal plane of P_1 is parallel to OQ_1 and that of P_2 parallel to OQ_2 as indicated.

A is the analysing nicol placed behind another aperture S_2, and is capable of rotation about a horizontal axis and its position can be read on a circular scale attached to it. The light coming out of the *A*-nicol is examined by the Galilean telescope *T*. The tube *D* containing the optically active solution is placed between S_1 and S_2.

Incident light rendered parallel by a lens *L* is made to fall upon the polarising nicol P_1. Since one half of the field of view is illuminated by light passing through the nicol P_1 and the other half, by light passing through both the nicol P_1 and – P_2 the field looks divided into two halves of unequal brightness separated by a sharp-line which is the image of the lower edge of nicol P_2.

If the principal plane of the *A-nicol* is perpendicular to the principal plane of the polarising nicol P_1 (*i.e.*, OQ_1), one-half of the field of view is bright and if it is perpendicular to that of P_2 (*i.e.*, OQ_2) the other half is bright. On the other hand, if the analysing nicol *A* has its principal plane perpendicular to *OR* — the intermediate position--both the halves are equally bright, so the direction of *OR* sets the zero position of the *A*-nicol. It is evident that for greater sensitiveness of the apparatus, θ should have a small value. In fact the values of θ will depend upon the brightness of the source and the transparency of the solution under examination. On the whole it may be remembered that for very small values of θ, the intensity becomes too low.

Polarimeter of Biquartz

A *biquartz* is another simple optical device which can be usefully employed for determining the rotation of the plane of polarisation. It consists of two semi-circular quartz plates *ACB* and *ADB* placed in juxtaposition each about 3.75 mm thick, one of right-handed (*R*) and the other of left-handed (*L*) quartz, cut with their optic axes perpendicular to their refracting faces. They are cemented together to form a circular plate, this biquartz plate is placed just in front of the polarising nicol like Laurent's half-shade plate. When plane-polarised white light is passed through each half of this plate its different components will be rotated through different angles by each half but in the opposite sense. For sodium tight this rotation is about 90°. Thus if *QOP* be the direction of the incident vibration, it will be rotated to the right in the plate *ACB* and the vibration of the yellow component will be parallel to *OQ'*, *i.e.*, at right angles to *OP*. In the other plate this component is rotated through 90° to the left and its vibration will be parallel to *OF*. Hence, the vibration of the yellow ray will be parallel to the same direction in both halves of the field, *viz.*, to the line *P'OQ'* perpendicular to *POQ*. Thus, if the principal plane of the analysing nicol be parallel to *POQ*, sodium light will be extinguished in both halves of the field and the other colours will be present in the same proportion in each half so that they will exhibit the same tint, a grayish violet, called the tint of passage.

A slight rotation of the *A* -nicol one way or the other will change this tint either to blue or red. On passing through this position the transition from red to blue is very rapid. Thus the zero position can be obtained with a fair degree of accuracy. When this is secured the plane of polarisation of the incident light is perpendicular to the principal plane of the analyser. Thus this arrangement even though designed for yellow sodium light is quite sensitive with white light for the proper setting of the instrument.

When the tint of passage is attained the substance under investigation is placed between the biquartz and the analysing nicol. If the substance rotates the plane of polarisation, the tint of passage will disappear from the field of the analyser making one-half red and the other half blue and to restore it the nicol must be rotated in the direction in which the plane has been rotated and through an equal angle.

Particular Rotation

To bring the rotation of all optically active substances into a comparable form the term specific rotation has been adopted.

'Specific rotation', at a given temperature t and for a given wavelength of light λ is defined as the rotation (in degrees) produced by a path of one decimetre length in a substance of unit density.

If θ is the rotation produced by l decimetre length of a solution of density d/cc, then the specific rotation S at a given temperature t and for a given wavelength of light λ used is given by

$$S_t\lambda = \frac{\theta}{I \times d} \quad ...(3)$$

i.e., specific rotation =

$$\frac{\text{rotation in degrees}}{\text{length in decimetres} \times \text{concentration in gm / cc.}}$$

Molecular rotation is given by the product of the specific rotation and the molecular weight of the substance. It is defined as the rotation produced by one decimetre length of solution containing one Gram-molecule of the active substance per cc of the solution.

Sugar is probably the most common of all the optically active substances and this optical activity is generally, employed for the estimation of its strength in a solution by measuring the rotation of the plane of polarisation. The instruments designed for determining the optical rotation produced by and substance are called polar-meters or saccharimeters. Essentially they consist of two nicols placed some distance apart and capable of rotation about the incident beam as axis. The nicols are first adjusted to be in the crossed position so that the field of view is dark. Then a tube containing the solution it introduced between the nicols, when it would be found that the field of view is no longer dark. To restore the position of extinction the analysing nicol it rotated through a certain angle θ demanding upon the length *of* the tube and the nature of the substance. Knowing θ, l and d the specific rotation is then calculated from the above, formula. We describe; below a few improved forms of polarimeters.

Polarisation: Rotatory Form

It is found that if plane polarised light be made to pass through certain substances the plane of polarisation of the emergent light is not the same as that of the incident light but has been rotated through a certain angle. This phenomenon is known as rotatory polarisation and this property of crystals and other substances is called optical activity and the substances which rotate the plane of polarisation are said to be *optically active.*

There are two types of optically active substances—those which produce clockwise rotation (when one looks towards the light source) are known as dextro-rotatory *(right handed)* substances-and others which rotate, the plane of polarisation in the anti-clockwise direction are called laevo-rotatory *(left-handed)* substances.

If monochromatic light from a source *S*, is made to pass through two nicols *PN* and *AN* put in a *conventional* crossed position, there is no emergent light. The polarising, nicol *PN* renders the beam plane-polarised with its vibrations in the principal plane. The analysing nicol *AN* in this position is set with its principal plane perpendicular to the direction of these vibrations; so that no fight is allowed to pass through it. If a quartz crystal cut *with its refracting faces perpendicular to the optic axis* is placed between the two nicols, still crossed, and light is incident normally on its faces, it is found that light is no longer cut off by the analysing nicol. To produce extinction of light the analysing nicol is required to be rotated through some angle. Thus the emergent light is still plane-polarised but its plane of polarisation has been rotated.

If the crystal is turned around so that the beam of tight passes through it in opposite direction, no change in the direction of rotation of the plane of polarisation is observed. Experiments have farther shown that some quartz crystals *are left-handed* and some right handed. Fused quartz is, however, optically inactive showing thereby that the optical activity is due to the crystalline structure of the substance.

Many liquids and organic substances in solution have been found to be optically active. Turpentine oil, aqueous solutions of various of sugars and of tartaric acid, etc., all cause the plane-polarised tight to be rotated through an angle which is proportional to the thickness of the medium which the light has to pass through.

The amount of optical rotation depends upon the thickness and density of the crystal or concentration in case of solutions, the temperature and the wavelength of light employed.

Optical Rotation: Explanation of Fresnel

Fresnel's explanation of optical rotation is based upon the assumption that when plane-polarised light is allowed to pass through a crystal *along the optic axis* it ft decomposed into two circulatory polarised vibrations rotating in opposite directions with the same frequency. In a crystal-like calcite which is not optically active, these two circular motions travel with the same speed and starting from the point of intersection of YY' with the circle arrive at the equidistant points, P_1 and P_2, at the same instant and in the same phase. Since any given point is their path is simultaneously arrived at it is obvious that at emergence they give rise to a linear vibration in the same plane as that of the incident vibration.

In optically active crystals these two circularly polarised vibrations travel with unequal speeds. As a consequence at emergence, the two circularly polarised; components have a certain phase difference between them, and combine; to form plane-polarised light whose direction of vibration will be inclined to the primitive plane of polarisation. The rotation of the plane of polarisation will thus be accounted for by the retardation of a circular component.

In a *right-handed quartz,* right-handed motion travels Caster than the *left-handed* while in a *left-handed quartz* left-handed motion travels faster than the right-handed. As a result of this the rotation of the vibration, plane is to the right or left according as the right-handed or the left-handed component is faster and is equal to half the phase difference on emergence between the two circular vibrations. These results may be arrived at very easily by means of the equations of the vibration. Since each circular motion is equivalent to two rectangular simple harmonic motions, so in a crystal which is not optically active the components of the right-handed circular motion beginning with the point of intersection of the circle with YY' axis are

$$x_1 = a \cos \theta = a \cos \omega t$$

$$y_1 = a \sin \theta = a \sin \omega t$$

The components of the left-handed circular motion are

$$x_2 = -a \cos \omega t, \; y_2 = a \sin \omega t$$

Thus the two circular motions, one right-handed and the other left-handed are equivalent to four simple harmonic motions whose resultant displacements along the x-axis and y-axis are

$$x = x_1 + x_2 = a \cos \omega t - a \cos \omega = 0$$

and

$$y = y_1 + y_1 = a \sin \omega t + a \sin \omega t$$

$$= 2a \sin \omega t.$$

Hence, the x components of the circular vibrations by the principle of superposition of wave motions cancel out and the resultant motion of the particle is simple harmonic being the sum of y-components of amplitude $2a$ executed along the Y-axis. This shows that the two oppositely circularly polarised vibrations rotating with equal frequency give rise to a plane-polarised vibration. The converse of this statement is equally true and has been also-verified experimentally by Fresnel with the help of his compound prism.

Now suppose the plane-polarised light passes through a dextrorotatory substance, then the right-handed circular motion being quicker, at emergence, starts from P_2 say, while the left-handed has not yet reached it but has attained a position P_1, being retarded by an angle ϕ from P_2. The equations of the transmitted component motions are

$$x_1 = a \cos \omega t;$$

$$y_1 = a \sin \omega t \qquad \text{... (right-handed)}$$

$$x_2 = - a \cos (\omega t + \phi);$$

$$y_2 = a \sin (\omega t + \phi) \qquad \text{... (left-handed)}$$

The resultant displacements along the two axes are:

$$x = x_1 + x_2 = a \cos (\omega t) - a \cos (\omega t + \phi t)$$

$$= 2a \sin (\omega t + \phi/2) \times \sin (\phi/2)$$

and

$$y = y_1 + y_2 = a \sin (\omega t) + a \sin (\omega t + \phi)$$

$$= 2a \sin (\omega t + \phi/2) \times \cos (\phi/2).$$

These two rectangular vibrations being in the same phase when compounded together give a resultant linear vibration along *CD* which makes an angle δ with the *Y-axis,* such that

$$\tan \delta = x/y = \tan (\phi/2).$$

i.e., $$\delta = \phi/2.$$

Thus the vibration plane of the incident plane-polarised light has been rotated in the clockwise direction by an angle which is equal to half of the phase difference between the two emergent circularly polarised vibrations.

We can prove the same relation for *a left-handed* substance.

Calculation of the Angle of Rotation: The magnitude of ft the angular rotation may be calculated in terms of the refractive indices of the right-handed and left-banded vibrations. If V_2 and V_1 be the velocities of the right-handed and left-handed vibrations, *t* the thickness of the crystal along the optic axis, the time retardation between the two circular components in traversing the right-handed quartz is

$$\Delta T = T_L - T_R = \frac{t}{V_L} - \frac{t}{V_R}.$$

If V_a be the velocity of light in air the path retardation

$$= V_a \Delta T = t\left[\frac{V_a}{V_L} - \frac{V_a}{V_R}\right] = (\mu_L - \mu_R)$$

where μ_L and μ_R are the refractive indices of the quartz for propagation along the optic axis of *L* and *R* circularly polarised component of the incident plane-polarised light.

Hence, the phase difference between the two circular components

$$\phi = \frac{2\pi}{\lambda} t (\mu_L - \mu_R)$$

∴ the angle of rotation of the plane of vibration by the right-handed quartz is given by

$$\delta = \frac{\phi}{2} = \frac{\pi t}{\lambda}(\mu_L - \mu_R). \quad \text{... (1)}$$

Similarly, the angle of rotation of the plane of vibration by the left-handed quartz is expressed by

$$\delta = \frac{\phi}{2} = \frac{\pi t}{\lambda}(\mu_R - \mu_L). \quad \text{... (2)}$$

Experimental Verification of Fresnel's Theory of Rotatory Polarisation: To verify his assumption experimentally Fresnel constructed a compound rectangular block made up of several right-handed and left-handed quartz prisms placed together alternately right and left. In each the optic axis was *parallel* to its respective base.

When a plane polarised beam is incident normally on the face *AB* of the block, it is decomposed into two component circular vibrations which are propagated along the optic axis with different speeds, right handed motion being faster than left-handed. On crossing *BC* the beam enters the left-handed quartz so that the right-, handed motion becomes slower and the left-handed faster. For the *R* component, therefore, the prism *L* behaves as a *denser* medium so by the ordinary law of refraction, this component is refracted towards the boundary normal and the *L* component away from it, for with respect to it the medium is *rarer*. At every boundary of the prisms the velocities are inter-changed. So that the ray bending towards the normal at one boundary bends away from it at the other. The net' result is that the angular separation of the two circular vibrations increases at each successive refraction and finally two distinctly separate beams emerge out of the block. On analysing the emergent beams with the usual device of a quarter-wave plate and a nicol, it is established that they are circularly polarised beams, one rotating to the right and the other to the left.

Instrumentations

Measuring the Spectrum

It is one of the most important optical instruments in a laboratory. It is used for the study of spectra produced by prisms and gratings, for the measurement of the dispersive powers, wavelengths of spectral lines, the refractive indices of solids and liquids, etc.

(a) A collimator C, consisting of a tube of variable length with an adjustable slit S at one end and a convex lens at the other. If a source of light, say, an arc lamp A is placed close to S which is at the focus of the convex lens, a parallel beam of light will emerge.

(b) A turn table R, provided with levelling screws on which a prism, a grating or any other optical device may be placed. It can be fixed at any desired height and can rotate about the central telescope T, provided generally with a Ramsden's eyepiece and cross-wires for examining the spectrum. It is also capable of rotation about a central vertical axis independent of the table. The position of the telescope is read on a graduated disc attached to it by means of two fixed verniers V_1 and V_2.

There are fixing as well as tangent screws for slow horizontal motion attached to both the table and the telescope.

Adjustments of a Spectrometer: Before using a spectrometer certain necessary adjustments have to be made.

1. *Alignment:* In the first place we must see the *alignment* of the axis of the telescope with that of the collimator so that they intersect the principal vertical axis of rotation of the telescope. This is generally done by the makers but may be checked by the student, if desired.
2. *Adjustment of the Telescope and Collimator for Parallel Light:*

Method 1.—The telescope alone is first directed towards a bright object, say, a lamp globe or the sky and the eyepiece is adjusted so that the cross-wires are as sharp as possible when seen through it. This fixes the position of the eyepiece relative to the cross-wires. Next pointing the telescope through an open window at some object a few hundred metres away such as a telegraph or electric post, the cross-wires and the eyepiece are racked together until a distinct image of the distant object is seen and there is no parallax of this image with that of the cross-wires.

The telescope is now turned towards the collimator with its slit illuminated with a monochromatic source of light, say, a sodium flame. The collimator slit is, then, adjusted until a distinct image of the slit falls on the cross-wires. The instrument is now adjusted so that parallel rays pass from the collimator to the telescope.

Method 2.— This method involves the use of a Gauss eyepiece and is used when a distant object is not easily available. It is a Ramsden's eyepiece with a thin plane plate of glass G, placed between its two components at an angle of 45° with the optical axis of the telescope. Light from a source S is admitted through an opening in the side of the draw tube and is reflected down the axis of the tube illuminating the cross-wires in its path. When the telescope is focused for parallel light and has its axis almost perpendicular to a plane reflecting surface placed on the prism table before the objective then an image of cross wires formed by rays reflected from the plane reflecting surface back into the objective will be at the principal focus of the objective. If this image coincides exactly with the cross-wires, the axis of the telescope is exactly normal to the reflecting surface. When the cross-wires and their images are both in sharp focus without parallax, the telescope is correctly adjusted for parallel light. For collimator adjustment proceed as suggested in method 1.

Method 3.— Schuster's method. In the absence of a distant object or a Gauss eyepiece, the following method due to Schuster may be used:

After focusing the eyepiece on the cross-wires, the telescope and the collimator are adjusted in a straight line. The slit is illuminated with sodium light and the prism is so placed on the prism table that it has maximum illumination from the collimator. The prism table is then rotated; slowly following the image with the telescope until the position of minimum deviation is found. The telescope is then moved through about 10° so as to enable it to receive images for two positions of the prism marked P and Q, with deviation greater than the minimum. Suppose the prism is in the position P. marked by full lines then the rays from the collimator fall on it more obliquely than for minimum deviation. The telescope is then focused' till the image is sharp and without parallax with the cross-wires. When the prism is next rotated to the position Q, the image will be found to be blurred. This time the adjustment of the slit be made with respect to the collimating lens till the image is in focus again. These alternatives of rotating the prism table and focusing first with telescope and then with collimator are repeated until turning the prism causes no change of focus. When this condition is achieved the rays from the slit are parallel in passing through the prism.

The Hilger's Constant Deviation Spectrometer: In this spectrometer, the telescope and the collimator C are fixed at right angles to each other on a cast iron stand. A constant deviation prism P is mounted on the prism table which can be rotated by a fine micrometer screw to which a drum D is attached. The prism may be regarded as made up of two 30° prisms and one right-angled prism. For any angle of incidence $i_{1,}$ the light of the wavelength which is at minimum deviation emerges at right angles to the direction of the incident beam after one internal reflection at the face BC. When the prism table is rotated, the spectrum; crosses the field and the drum passes under an index which indicates the wavelength of any spectral line which fails on the cross-wires.

The instrument can serve as a monochromatic illuminator. For this purpose it is necessary only to remove the eyepiece and place a writable aperture in the focal plane of the spectrum.

Spectroscope and Spectrograph: If a spectrometer is merely designed for the qualitative study of the spectra, that is, to view the various spectral lines emitted by a source, it is called a spectroscope. One convenient type, which gives a spectrum with practically no deviation is called a direct vision spectroscope.

If it is desired to have a permanent record of the spectrum of a source, the eyepiece of the telescope is removed and a photographic plate is placed in the focal plane of the objective. Such an instruments is then called a spectrogram. By noting the position of a line, or the separation between lines of different wavelengths we can find the wavelengths of the spectral lines if that of a particular line is known. These instruments are now commonly used with great success in the study of spectra especially in the ultraviolet region and the near infrared region as such regions cannot be detected by the human eye.

For photographing these regions plates with specially prepared emulsions are used. The lenses and prisms are made of quartz as glass absorbs the light from these regions.

Different Eyepieces

The Eyepiece: If a single convex lens is used as an eyepiece, besides the effects of chromatic and spherical aberrations it suffers from a defect arising from an entirely different cause. When the eye is placed close to the eye lens, the rays from the peripheral parts of the image *I* of a distant object formed by the objective *O* would cross over the eye lens and thus fail to enter the eye along with the paraxial rays. Thus the plane of the field near the axis will alone be seen and the field of view would become limited.

Hence to enable the marginal rays to pass through the eye-lens or in other words to increase the field of view another lens *F* is introduced between the eye lens and the objective, say, in the plane of the image *I*. This deflects the outer rays towards the principal axis so that after passing through the eye lens these rays enter the pupil of the eye simultaneously with the central rays thus enlarging the field of view. *F* is called the *field lens, F improves the field* of view and *E* the *eye lens, F* improves the field of view. And *E* acts as a magnifier. The combination of *E* and *F* is termed the **compound eyepiece** or **ocular.** The two lenses are so designed that the combination is free from spherical and chromatic aberrations. There are several types of oculars used with various optical instruments. They consist of lenses of appropriate focal lengths and radii of curvature and are placed at a suitable distance from each other.

Kellner's Eyepiece: It consists of two plano-convex lenses of equal focal length and separated by a distance equal to the focal length of either. They are placed with their convex surfaces turned towards the incident light. The condition for achromatism.

$$a = \frac{f_1 + f_2}{2} \text{ is fully satisfied.}$$

Since $f_1 = f_2 = f$, say, so that $a = \frac{f + f}{2} = f$.

the focal length of either. This condition simply means that red and blue images although of different sizes and formed at different positions subtend the same angle at the eve causing the sensation of whiteness due to overlapping. Spherical aberration is also reduced as the deviation is now spread over four refracting surfaces. In order that the rays must emerge from the eye lens E as a parallel beam for the final image to be seen at infinity before striking E they must have come from its first focal plane. But as the distance between the field lens and eye lens is f_r the focal length of either, the principal plane of the field lens coincides with the first focal plane of the eye lens. Hence the objective must form a real image in the principal plane of the field lens. This will make no difference in the dimensions of the image finally seen since no further magnification is produced by the field lens.

Merits and Demerits

(1) This eyepiece has a very wide field of view and is suitable for use with microscopes.

(2) Since the magnification is entirely produced by the eye lens the marginal parts are disproportionately magnified thus producing distortion.

(3) There is another undesirable feature that the image I and the surface of the field lens are simultaneously in focus, so that any dust particles or a smear on the surface of F or any flaw in its material would be magnified by the eye glass and would seriously interfere with the visibility of the image.

Ramsden's Eyepiece: This is generally used in telescopes and spectrometers and is designed to enable us to measure the dimensions of the image by the aid of a scale or cross-wires in the eyepiece. It consists of a combination of two plano-convex lenses of equal focal length separated by a distance two-thirds of their common focal length and placed beyond the image formed by the objective. The lenses are placed with their plane faces outwards so that spherical aberration effects are reduced. Since the distance apart of the lenses is only $\frac{2}{3}$ of their common focal length instead of f on account of the disadvantage referred to in Kellner's eyepiece, the condition for achromatism is not quite satisfied but the departure from perfect achromatism is not large. Chromatism is sometimes further reduced by using an achromatic lens for each component.

The spherical aberration is reduced by sharing the deviations as equally as possible among the four refracting surfaces. This also eliminates coma to some extent. As narrow pencils pass through the eyepiece, this also helps in reducing the radius of the circle of least confusion due to spherical aberration and astigmatism.

The course of rays through a Ramsden's eyepiece. To examine the image without strain the rays leaving E must be parallel and consequently before striking E must have proceeded from A_1 B_1 in its focal plane. A_1B_1 is the image formed by the field lens of the real image AB formed by the object glass at its second focus.

$\therefore$ $EF = \frac{2}{3} f$, the distance from F to $A_1B_1 = \frac{f}{3}$ numerically,

i.e., $v = -\frac{f}{3}$ for the field lens.

Since $$\frac{1}{v} - \frac{1}{u} = \frac{1}{f},$$

$$\therefore \qquad -\frac{3}{f} - \frac{1}{u} = \frac{1}{f},$$

whence $$\frac{1}{u} = -\frac{3}{f} - \frac{1}{f} = -\frac{4}{f}, \text{ or } u = -\frac{f}{4}.$$

That is, the real image AB is in front of the field lens. This is the position at which cross-wires must be placed and at which real image due to objective is formed. Hence the eyepiece is sometimes known as positive eyepiece. It magnifies equally both the image at AB and the cross-wires or the transparent micrometer scale placed there. Therefore, it can be easily focused and enables us to take accurate measurements of the size of the image and there is very little distortion in it.

The equivalent focal length F' of the compound eyepiece is given by

$$\frac{1}{F} = \frac{1}{f_1} + \frac{1}{f_2} - \frac{a}{f_1 f_2}$$

$$= \frac{1}{f} + \frac{1}{f} - \frac{2f/3}{f^2} = \frac{2}{f} - \frac{2}{3f} = \frac{4}{3f}.$$

$\therefore$ $$F' = 3f/4.$$

The equivalent lens of this focal length should be placed *f/2 behind* the field lens.

The magnifying power of a telescope fitted with Ramsden's eyepiece is increased somewhat as

$$M = \frac{\text{focal length of objective}}{\text{focal lenght of eye - piece}}$$

$\therefore$ $$M = \frac{F}{F'} = \frac{F}{3f/4} = \frac{4}{3} \cdot \frac{F}{f},$$

since the equivalent focal length F' of eyepiece is $3f/4$.

Cardinal Points: The focal length of the equivalent lens and the position of the principal foci and the unit planes may also be determined by the use of equations (2), (3) and (4).

$$F = +\frac{f_1 f_2}{f_1 + f_2 - a} = \frac{f^2}{2f - \frac{2}{3}f} = +\frac{3}{4}f,$$

$$\beta = -\frac{f_1 f_2}{f_1 + f_2 - a} = -\frac{f \cdot \frac{2}{3}f}{2f - \frac{2}{3}f} = -\frac{f}{2}$$

and $$\alpha = +\frac{f_1\, a}{f_1 + f_2 - a} = -\frac{f \cdot \frac{2}{3}f}{2f - \frac{2}{3}f} = +\frac{f}{2}.$$

The positions of the principal points H_1 and H_2 and the principal foci F_1 and F_2 are marked with the aid of the above values. Since the system is in air, the nodal points coincide with the principal points. From this we can easily find that the distance of the first principal focus from the field lens of the eyepiece is given by, $F_1 A = F_1 H_1 - \alpha$

$$= \frac{3f}{4} - \frac{f}{2} = \frac{f}{4}$$

Similarly, the distance of the second principal focus from the eye-lens,

$$BF_2 = H_2F_2 - \beta = \frac{3f}{4} - \frac{f}{2} = \frac{f}{4}.$$

For Guass eyepiece which is a modification of Ramsden's eyepiece and is used in spectrometric measurements.

Huygens' Eyepiece: This eyepiece is so designed that when two converging lenses are used the condition for partial achromatism is satisfied and the spherical aberration is *minimum*. For the combination of two thin lenses to be achromatic, the distance apart of the lenses should be half the sum of the focal lengths of the component lenses.

viz., $$a = \frac{f_1 + f_2}{2} \qquad ...(i)$$

and for minimum spherical aberration

$$a = f_1 - f_2 \qquad ...(ii)$$

Combining (*i*) and (*ii*), we get

$$f_1 - f_2 = \frac{f_1 + f_2}{2}, \text{ or } f_1 = 3f_2$$

and $$a = 3f_2 - f_2 = 2f_2.$$

The eyepiece, therefore, consists of two thin lenses so that the focal length of the field lens f_1 is 3 times the focal length of the eye lens f_2, and the distance apart twice the focal length of the eye lens. Both the lenses are plano-convex of the same kind of glass arranged with their convex faces towards the incident rays in order to reduce the various aberrations.

To trace the pencil of rays through a telescope fitted with Huygens' eyepiece, we observe that in order that the rays must emerge from the eye lens as a parallel beam, the field lens must form an image A_1B_1, at the first focal plane of the eye lens, *i.e.*, midway between the two lenses. Therefore, if AB is the position of the image formed by the object glass, then applying the relation

$$\frac{1}{v} - \frac{1}{u} = \frac{1}{f_1}$$

to the field lens, we have

$$\frac{1}{f_1} - \frac{1}{u} = \frac{1}{3f_2},$$

Since $\qquad v = f_2$

and $\qquad f_1 = 3f_2.$

$$\therefore \qquad \frac{1}{u} = \frac{1}{f_2} - \frac{1}{3f_2} = \frac{3-1}{3f_2} = \frac{2}{3f_2}$$

or $\qquad u = \dfrac{3f_2}{2}$

This shows that the image AB is *virtual* and it is on the same side of the field lens A_1B_1. Hence a parallel beam falling on the objective converges towards AB but is intercepted by the field lens and brought to focus at A_1B_1. Proceeding these rays strike the eye glass and emerge as a parallel beam.

As the rays from the objective are deviated by the field lens before the image is formed, the first focal plane of the objective is virtual and no cross-wires or a micrometer scale can be used in this eyepiece in the focal plane of the objective. The eyepiece is, therefore, known as a negative eyepiece and is not well-adapted for telescopes. But if cross-wires are required to be used with this eyepiece, they must be placed midway between the eye and the field lenses, *i.e.*, in the position A_1B_1. Accurate measurements, however, would not be possible as the cross-wires, or the scale would be viewed by the rays which have traversed through the eye lens only, while the image would be seen by rays that have passed through both the lenses of the eyepiece and, therefore, except close to the centre on the field of view the two images would suffer from different amounts of aberration and distortion being disproportionately magnified. The eyepiece also suffers from considerable curvature of the field, the curvature being convex towards observer's eye. The field of view of this eyepiece is also inferior to the Ramsden's eyepiece. It was to avoid these defects that Ramsden's eyepiece was designed.

The equivalent focal length F' of this eyepiece is given by

$$\frac{1}{F'} = \frac{1}{f_1} + \frac{1}{f_2} - \frac{a}{f_1 f_2} = \frac{1}{3f} + \frac{1}{f} - \frac{2f}{3f^2}$$

$$= \frac{1}{3f} + \frac{1}{f} - \frac{2}{3f} = \frac{1+3-2}{3f} = \frac{2}{3f}$$

$$F' = 3f / 2.$$

It must be placed a distance f behind the eye lens to render the rays from AB parallel.

The magnifying power of a telescope fitted with Huygens' eyepiece is somewhat reduced,

$$M = \frac{F}{F'} = \frac{F}{3f/2} = \frac{2}{3}.\frac{F}{f},$$

where F is the focal length of the objective.

Cardinal Points: We can also study the behaviour of a Huygens' eyepiece by determining the focal length of the equivalent lens, the positions of the principal points and first and second principal foci.

Let $f_2 = f$, say, then $f_1 = 3f_2 = 3f$.

Now $$F' = +\frac{f_1 f_2}{f_1 + f_2 - a} = +\frac{f \times 3f}{4f - 2f} = \frac{3f^2}{2f} = \frac{3}{2} f,$$

$$\beta = -\frac{f_2 a}{f_1 + f_2 - a} = -\frac{f \times 2f}{4f - 2f} = -\frac{f \times 2f}{2f} = -f,$$

$$\alpha = +\frac{f_2 a}{f_1 + f_2 - a} = +\frac{3f \times 2f}{f_1 + f_2 - a} = \frac{6f^2}{2f} = 3f.$$

Using these values the positions of the principal points H_1 and H_2 and the principal foci F_1 and F_2 *are* marked. Since the system is in air, the nodal points coincide with the principal points. The first and second principal foci are formed at a distance points of $f/2$ in front of and behind the eye lens and the principal or unit planes are at distance f behind and in front of the eye lens.

Comparison of eyepieces

Ramsden's eyepiece	*Huygens' eyepiece*
1. The image formed by the objective is real and lies in front of the field lens at a distance $f/4$ from it. This is the position at which the cross-wires must be placed. The cross-wires and the image are equally magnified. The eyepiece is sometimes known as positive eyepiece.	1. The image formed by the objective lies behind the field lens, midway between the two lenses. No cross-wires can be used. But if cross-wires are required to be used, the cross-wires and the image will suffer from different amounts of aberration and distortion being disproportionately magnified. The eyepiece is, therefore, known as negative eyepiece.
2. It does not satisfy the condition for minimum spherical aberration.	2. It satisfies the condition for minimum spherical aberration, $a = (fi - f_2)$.
3. It does not satisfy the condition for achromatism. However, it can be made achromatic by the use of achromatic doublets.	3. It satisfies the condition for achromatism, $a = (f_1 + f_2)/2$.

Contd...

Ramsden's eyepiece	*Huygens' eyepiece*
4. The image formed is almost flat, *i.e.*, other aberrations are better removed.	4. The image formed is slightly convex towards the eye, *i.e.*, other aberrations such as distortion, etc. are not well removed.
5. Eye-relief* is 50% greater in this case.	5. Eye-relief is too small.
6. It is used in telescopes and other optical instruments where micrometer scales are used for accurate measurements. Since it possesses a flatter field, hence scales covering the entire field can be employed.	6. It is generally used in microscopes and other optical instruments for qualitative purposes only.
7. It can be used as a simple microscope, since in this first principal focus lies in front of the field lens.	7. It cannot be used as a simple microscope, because the first principal focus in this case lies between the field lens and the eye lens and is virtual.

Advanced Microscope

This is an optical arrangement for viewing extremely minute particles of matter, say, some colloidal suspension although no image of the particles is formed and their size and shape, etc. cannot be ascertained. The suspension to be examined is contained in a vessel *V* with an optically worked glass window on one side and at the top. In this case the illumination is not direct as in the compound microscope, but a very thin beam of intense light from an arc lamp *S* is allowed to enter the solution horizontally through the side glass window in *V*. It is brought to Principle of ultra-focus on the axis of the microscope, scope *M* by a convex lens *L*. The suspended particles falling in the path of the incident beam scatter light and when focused from above by the microscope held vertically they can be observed only by scattered light against a comparatively dark background, called the dark field illumination.

In this way even those suspended particles can be seen which are far too small to be made visible by ordinary means of illumination. Indeed we can see down to those of which the diameter is 10^5 mm or 1/10 of the wavelength of sodium light. In order that particles should scatter enough light in a direction normal to the incident light, besides intense illumination there should be sufficient difference in the optical refractivities of the liquid and the suspension. To study details ultraviolet light is employed.

By this arrangement Brownian movements can be easily examined. A small metallic chamber containing smoke has a high power convex lens on one side and an optically worked glass window at the top. Parallel light from an arc lamp is focused by the side lens to a point in the chamber on the axis of the microscope very close to its plane glass top through which the scattered light is viewed by means of the microscope. Very queer quivering movements of the smoke particles shining like starry clusters in the dark sky are easily visible.

These movements are due to the random kicks received by the visible smoke particles from the invisible air particles and are a striking confirmation of the kinetic theory of gases. They are known as Brownian movements because they were originally observed by the British botanist Dr. Robert Brown in the study of the plants.

There are alternative arrangements for obtaining dark field illumination. In these types the light is incident upon the object at angles such that it does Not pass by transmission or ordinary specular reflection directly into the objective, that is, the only light to enter must be that which has been diffractor scattered by the object itself. For low power objectives this is achieved by inserting below the substage condenser a circular opaque disc with its centre on the axis of the microscope; its diameter is such that it completely obstructs the central rays which would otherwise enter the objective.

For high power oil immersion objective the improved type of dark field illumination is of the reflecting type which makes use of a cardioid lens. The illuminating rays are reflected first at the spherical reflector *S* and then at the surface of the cardioid *C* and fall on the object at much greater obliquity than in the previous case. By this device both chromatic and spherical aberrations are removed and the condenser becomes truly aplanatic.

Defects of Objectives and Eyepieces and their Removal : In discussing the simple theory of a telescope and a microscope single lenses were used as objectives and eyepieces but we have noticed already that such lenses suffer from various defects of spherical and chromatic aberrations and a limited field of view. Hence in actual practice instead of using a single lens for each a suitable combination of lenses is employed.

Objectives:

(i) *Telescope Objective:* We have described how the chromatic aberration of a telescopic objective can be minimised by using combination of a convergent lens of crown glass and a divergent lens of flint glass cemented together by canada balsam. The combination is as a whole plano-convex with convex side facing the parallel rays from the distant object so as to remove spherical aberration.

(ii) *Microscope Objectives:* High quality microscopes are equipped with short focus objectives consisting of coaxial lens systems which make them free from chromatic and spherical aberrations even when a large aperture is used to admit a great deal of light. The necessary corrections for spherical aberration, etc. are made by making use of aplanatic surfaces and those due to chromatism by suitable achromats of crown and flint shows Abbe's homogeneous immersion objective. The first lens L_1 of the objective is a pianohemispherical convex lens and the space between this and the object to be examined is filled with cedar wood oil of the same refractive index as that of glass. The advantage of this method of *immersion* is that the divergence of rays emerging from glass side into air is eliminated so that the objective receives more rays from the object than it would, if oil were not used. For in the absence of the oil some of the rays which strike the objective when the oil is present, emerging from the side into the air would be bent outwards so far as to miss the objective.

A wide angled cone of rays proceeding from the object. Abbe's homogeneous A which is one of the aplanatic points immersion objective of surface S of $L_{1,}$ after refraction from S will appear to proceed from A_1. A second lens L_2 is placed above it so that A_1 is the centre of curvature of its lower surface $S_{1.}$

Thus the rays are refracted only at the second surface S_2 of L_2 It is also arranged that A_1 and A_2 are aplanatic points with respect to this surface. We have thus a wide-angled beam originating from A being converted into a beam of smaller angle coming from A_2 and making its entry into the microscope.

In order to correct for chromatic aberration two compound lenses L_3, and L_4 of crown and flint glasses are placed above L_2. Such a system of lenses satisfies Abbe's sine condition so spherical aberration and coma are both removed even when a wide-angled pencil of rays is used. The achromatic doublets produce a colourless image in front of the eyepiece.

Complex Microscope

The subject of microscopy is gaining more and more importance in recent times and greater magnifications and resolving powers have become necessary. In modern work, microscopes, with electronic beam as a source of light, are being employed, thus improving the technique of microscopy considerably. We shall now discuss the principle of a simple form of compound microscope. To obtain very large magnifications and get rid of the several system a compound microscope is generally employed for the study of minute substances.

In its simplest form a compound microscope, consists of two convergent lenses, one of very short focus called the objective *O* and the other of somewhat longer focus called the eyepiece *E*. The object *AB is* situated just outside the focus of the objective which forms its real image at *ab*. This image becomes the object for the eyepiece which acting as a magnifying glass forms a large virtual image at *A'B'* anywhere between the near and the far points of the eye. The overall magnifying power *M* of the instrument will clearly be given by the product of linear magnification *Me* due to the objective and the magnifying power *M,* of the eyepiece,

i.e., $$M = M_0 \times M_{e.}$$

The linear magnification due to the objective is given by

$$M = \frac{v}{u},$$

where v is the distance of *ab* from *O*,

$$\therefore \qquad M = \frac{v}{u}\left(1 + \frac{D}{f_e}\right) \ldots [\because M_e = 1 + \frac{D}{f_0}]$$

Also $\frac{v}{u} = 1 - \frac{v}{f_o}$, for the objective, so that when f_0 is small, this may approximately be taken as $-v/f_o$. The eyepiece of focal length *fe* acting as a magnifying glass, gives the magnification M_1 approximately equal to D/f_e. Hence dropping negative sign, we have

$$M = M_0 \times M_e = \frac{v}{f_0} \times \frac{D}{f_e}$$

The system of lenses generally comprising the objective of a microscope is a very complex one and replacing it by principal planes as the value of M_e may be written as $\frac{L}{f_0}$ where *L* is the distance bF_2 of the image ab from F_2, the second focal point of the objective and $f_o = H_2F_2$, its focal length.

Hence $$M = \frac{L}{f_0} \cdot \frac{D}{f_e} \qquad \text{...(10)}$$

L is called the optical length of the microscope and is generally taken as 18 cm. Since D for normal eye =25 cm,

$$M = \frac{18 \times 25}{f_0 \times f_e}$$

$$= \frac{450}{f_0 \times f_e} \qquad \text{... (11)}$$

It should be noted that this expression for magnifying power is independent of the nearness of the observer to the eye lens.

In practice the objectives are specified by their focal lengths, 1. 27 cm ($=\frac{1}{2}''$) objective and 0.635cm($=\frac{1}{4}''$) objective, etc., and the eyepieces by their magnifying power so that the magnifying power of the microscope can be readily calculated.

The optical arrangement of a compound microscope generally employed for examining small particles or slides. It is provided with a rotating nose-piece to which are attached three objectives of different magnifications, one of them being the oil immersion objective. By rotating the nose-piece, any one of the objectives may be set in proper alignment with the eyepiece.

The object or the slide to be examined is placed on the stage S of the microscope with a hole in the centre to allow light to pass through it. Immediately below the stage is a thick convex lens C called the condenser. Below this is a mirror M, plane on one side and concave on the other, capable of rotating about a horizontal axis. Light rays from a strong source of light are reflected by the mirror so as to fall on the condenser, which in turn focuses them on the object to be examined.

The Exit Popil and Magnifying Power: As in the case of a telescope, the exit pupil of a microscope is the image of the aperture of the objective formed by the eye lens and it is the place where the eye must be placed to see the greatest *range* of the object. From the course of rays it is clear that *KJ* represents the exit pupil of the microscope. If *r* be its radius, then

$$2r = KJ = MN = 2f. \tan \alpha_{1,} \qquad ...(a)$$

where f_e is the focal length of the eye lens and

$$a_1 = \angle PDO = \angle NDE.$$

Since $\angle PCO(=\alpha)$ is the object slope angle for the objective and α_1 is the corresponding image slope angle, then by the sine condition we have

$$\mu h \sin \alpha = \mu_1 h_1 \sin \alpha_1,$$

where h_1 is the linear length of the image formed by the objective of a small axial object of corresponding length *h*.

It follows that M_o, the magnification due to the objective is given by

$$M_0 = \frac{h_1}{h} = \frac{\mu \sin \alpha}{\mu_1 \sin \alpha_1} \qquad ...(b)$$

The quantity μ sin α, that is, the product of the refractive index of the object medium and sine of the slope angle of the outermost ray from an axial point on the object is called the numerical aperture (*N.A.*) of the objective of a microscope. The name was given by Abbe (1840—1905). It is important chiefly in connection with the brightness of image, depth of focus and the resolving power of the microscope.

Now equation (*b*) may be written as

$$M_o = \frac{N.A.}{\sin \alpha_1}, \text{ since } \mu_1 = 1.$$

As α_1 is small we may put $\sin \alpha_1 = \tan \alpha_1$.

$$\therefore \qquad M_o = \frac{N.A.}{\tan \alpha_1}.$$

Since the magnifying power of the microscope.

$$M = M_o \times M_o = M_o \times \frac{25}{f_e},$$

$$\therefore \quad M = \frac{N.A.}{\tan \alpha_1} \times \frac{25}{f_e} = \frac{25\, N.A.}{r} \quad \ldots$$

[from (*a*) above] ...(12)

$\therefore$ r, the radius of the exit pupil $= \dfrac{25\, N.A.}{M}$ cm ... (13)

Now, for an eye examining small objects in average illumination the diameter of the pupil is 2.5 mm and its radius=l.25 mm. If the optical system is such that the pupil of the eye is completely filled With light emerging from the eyepiece, the magnifying power is said to be normal.

$$\therefore \quad \text{normal } M = \frac{25\, N.A.}{0.125} = 200\, N.A. \quad \ldots(14)$$

In actual working, however, a magnifying power much greater than the normal is needed even at the cost of reducing the diameter of the eye-ring as will be explained in the discussion of the resolving power of a microscope. If magnifying power is less than the normal, the radius of the exit pupil exceeds that of the eye pupil and the whole beam of light is not utilised. In this case the actual brightness of the images remains the same as for normal magnifying power but the numerical aperture which is actually used is reduced and there is, as we shall see later, a consequent loss in the resolving power.

Now let us find the magnifying power of a compound microscope when the observer's eye is at the exit pupil and the image is at the least distance of distinct vision.

Let E_1 be the aperture of the objective and E_2 the image of this aperture produced by the eyepiece, *viz.*, the exit pupil. All the rays from the object which enter the microscope must pass through E_2. Hence, if an eye is placed close to the exit pupil with its pupil larger in size than E_2, the image is seen under the best conditions.

Let p, q and x be the distances as indicated and v be the distance of the final image from the eye lens then when $v = D - q$, the magnifying power is given by

$$M = \frac{h_2}{D} \div \frac{h_2}{D} = \frac{h_2}{h_1} \times \frac{h}{h_1}$$

$$= \frac{D-q}{p-x} \times \frac{f_o}{u+f_o}.$$

Since $\frac{1}{q} + \frac{1}{p} = \frac{1}{f_e}$ and also $-\frac{1}{D-q} + \frac{1}{p-x} = \frac{1}{f_e}$, we have

$$\frac{p+q}{pq} = \frac{1}{f_e} \text{ and } \frac{D-q}{p-x} = 1 + \frac{(D-q)(p+q)}{pq}.$$

$$M = \left[1 + \frac{(D-q)(p+q)}{pq}\right]\left(\frac{f_o}{u+f_o}\right). \qquad \text{...(15)}$$

Operative Camera

Photographic Camera: The principle of a photographic camera is that of a convex lens forming a real image on a sensitive plate held at the back of a light-tight chamber formed by canvas bellows with blackened interior. The lens is mounted on a slide so that its distance from the sensitive plate can be altered for bringing to sharp focus the images of objects situated at varying distances.

An adjustable stop is placed in front of the lens so that the amount of light passing through it may be varied. It can be easily shown that the intensity of illumination E of the image is directly proportional to the area of the aperture and inversely proportional to the square of the focal length of the lens, *viz.'*

$$E \propto \frac{\text{area of the aperture}}{(\text{focal length})^2}$$

If d be the diameter of the aperture and f the focal length of the lens, then

$$E \propto \frac{\pi\left(\frac{d}{2}\right)^2}{f^2} = \text{constant} \times \frac{d^2}{f^2}. \qquad \text{... (1)}$$

The quantity d/f is called the aperture ratio of the lens. It is clear from equation (1) that different lenses with the same value of d/f will give the same brightness of a given object.

The reciprocal f/d of the aperture ratio is called the focal ratio or f number* of the lens. Thus, a lens of 20 cm focal length and 4 cm diameter is said to have a f-number of 5 which means that the diameter of the lens aperture is $f/5$.

Since exposures needed under given conditions may be assumed to be proportional to the square of this number, the apertures are usually indicated in a series such that the exposures required are doubled in passing from one to the next, thus we have the series

$$\frac{f}{2}, \frac{f}{2.8}, \frac{f}{4}, \frac{f}{5.6}, \frac{f}{8}, \frac{f}{11.3}, \frac{f}{16}, \text{ etc.}$$

Speed of a Lens and Depth of Field: The speed of a lens is the measure of the rapidity with which the photographic image is built up. It is obviously proportional to the gain of brightness which as we have seen above is given by

$$E = \text{constant} \times \left(\frac{d}{f}\right)^2 = \frac{\text{constant}}{(f/d)^2} = \frac{\text{constant}}{(f-\text{number})^2}$$

Hence the speed of a lens varies inversely as the square of the f-number. For instance, an $f/2$ lens is faster than $f/4$ lens in the ratio and is more suitable for taking the photographs of faintly illuminated objects and of those which are moving relatively to the camera and hence require a very short exposure.

$$\left(\frac{f}{2}\right)^2 \div \left(\frac{f}{4}\right)^2 = 4:1$$

But it is limit worth noting that with a lens of wide can relative aperture, besides the cumbersome aberrations, points like *O* not accurately in focus give comparatively large circles of confusion like *bb* on the plate P placed. The range of object distances giving reasonably clear images will be small and this is expressed as the lens having a little depth of field.

It is clear that if the lens aperture is reduced then for a given position of *O* the diameter of the circle on the plate will become smaller thus increasing the depth of field, that is, the range of object distances besides improving the definition of accurately focused images. This explains why in a simple box camera with a small aperture and the film placed in the focal plane of the lens, objects very close to the camera lens can produce a reasonably clear picture.

Photographic Objectives: The most important feature of a good camera lens as a rule is that is has a high relative aperture with a large field of view and suffering as it does from the defects of overcome in its construction. The designers of such objectives have therefore, resorted to compromises as regards these corrections that best serve their purpose.

Three types of lenses are in general use:

1. Meniscus lens
2. Rapid rectilinear
3. Triplet anastigmats: Zeiss Tessar.

The above two types of objectives have largely been superseded by recent triplet anastigmats, the most prominent of them being Zeiss Tessar. The chief quality of these lenses is that they give a flat field by actually removing the defects of astigmatism and curvature by making the power of the central flint element equal to the sum of the powers of the crown elements. By placing the concave lens between the two convex ones, the marginal rays can be made to pass through the negative lens so close to the axis that on the whole the triplet is of positive power.

This is a very fast lens and is generally used in the motion picture cameras.

Telephoto Lens: The size of the image of a distant object is proportional to the angle which the object subtends at the camera lens and the focal length of the lens. Hence to obtain a fair-sized image of some distant object, an objective of large focal length is required but this would make the camera itself inconveniently long and unmanageable.

Therefore, photographs of distant scenes are taken with a telephoto lens. It consists of two achromatic lenses, one being convex and the other concave. The latter occupies the normal lens position in the camera box while the former (convex) is placed some distance in front of it. The distance between them is less then the focal length of the convex lens L_1. The chief advantage lies in the fact that while the focal length is long the distance from the back of the objective to its second principal focus is such as can be easily accommodated in a folding camera.

Parallel rays from a very distant object incident on L_1 are brought to focus at F behind the first focus of the divergent lens which is of short focal length. The point F serves as a virtual object for the concave lens which brings them to focus at F_1 where the sensitive plate is placed. If these rays are produced backwards until they cut original beam at P, then it is seen that the combination is equivalent to a single convex lens placed at PN, its focus being at F_1 Without the divergent lens the length of the camera would be approximately FA; the divergent lens increases it to F_1A but gives a greatly increased focal length to the combination for the power is decreased to

$$\frac{1}{f_1}+\frac{1}{f_2}-\frac{a}{f_1 f_2},$$

where f_1 and f_2 are the focal lengths of the component lenses and a, the distance between them. Thus the telephotographic camera has an effective length which is much greater than its actual length. Such a system of lenses then enables us to take highly magnified pictures with an ordinary camera.

The focal length of a telephoto lens can be varied at will by altering the distance between its two components. But the advantage thus gained will result to loss of definition for settings other than the standard one for which necessary corrections have been made. Hence it is usually desirable to have telephoto objectives with a fixed focal length.

The Sextant: This instrument was devised by Hadley for measuring the angular separation between two distant objects or the altitude of heavenly bodies. It is based on the principle that *the deviation produced in a ray by successive reflections from two inclined mirrors is constant for all angles of incidence and is twice the angle between the mirrors.*

It consists of a rigid triangular frame being the sector of a circle having an angle of 60°, *i.e.,* one sixth part of the circumference and hence the instrument is called a sextant. There are two plane mirrors *A* and *B,* set at right angles to the plane of the frame. The mirror *A* called the index glass, is the movable mirror and is placed at the centre of the circle. It is fixed to an arm *AV* which moves over the *graduated* arc *PQ* and carries a vernier scale *V* capable of reading upto 10 to 15 seconds. The mirror *B,* called the horizon glass is, fixed to one of the rigid arms *AP.* It is half silvered so that if an object is viewed through the telescope *T* attached to the other arm *AQ* and directed towards *B,* two images are in general observed, one seen directly along *CBE* and the other seen after two reflections along the path *DABE,* when the movable arm is on the zero of the scale, so that the two mirrors are parallel to each other. In this position a distant object is seen simultaneously, the two images exactly coinciding with each other.

To measure the angle subtended at the position of the sextant by two distant points in a vertical line, the sextant is set in a stand with its plane vertical so that the lower of the two points is at the same height as A. The telescope is directed to see the lower point through the unsilvered portion of B and the movable arm is then rotated until the image of the upper point H, seen through the telescope, after two reflections at A and B coincide with that of the lower point. Suppose $HABE$ is the path of rays from the upper point. Let NAM and BM, the normals to mirrors A and B in this position meet in M then the angle AMB (= a) between them, is equal to the angle AFB between the mirrors. Now by the law of reflection,

$$\angle HAN = \angle NAB = \theta \text{ (say)}$$

and

$$\angle ABM = \angle MBE = \phi \text{ (say)}.$$

If β is the angle which the two points subtend at E, we have from Δ AEB, the exterior

$$\angle HAB = \angle ABE + \angle AEB,$$

or

$$2\theta = 2\phi + \beta \qquad \text{...}(i)$$

Similarly, from Δ ABM,

$$\theta = \phi + \alpha,$$

where α is the angle between the mirrors A and B and hence also between their normals.

From (*i*) and (*ii*) $\beta = 2\alpha$, which means that the angle which the two points subtend at the observer's eye at E is double the angle between the mirrors A and B.

When $\alpha = 0$, $\beta = 0$, that is, A and B are parallel and the direction of the arm A V is then along AQ parallel to the fixed direction BF of the horizon glass. If the zero of the main scale coincides with the zero of the vernier scale in this position, α is read directly on the arc. Actually, however, the graduations on the arc PQ are doubled. That is instead of 60 divisions of 1° each there are 120 divisions so that instead of a, it is the required angle $\beta = 2\alpha$, which is read directly on the scale.

Determination of the Height of a Distant Object: The angle which the distant object AB, say a pole, subtends at a point C at the same level as the foot *B* is determined with the help of the sextant. If *h* is the required height and z (= $x + y$) the distance of the pole from C and θ_1 the angle subtended by *h* at C, then $h = z \tan \theta_1$.

If the object is inaccessible, the angles θ_1 and θ_2 which it subtends at two different positions C and *D* distance *x* apart, are determined. Then we have

$$X + y = h \cot \theta_1 \qquad ...(iii)$$

and

$$y = h \cot \theta_2 \qquad ...(iv)$$

On subtracting (*i*) from (*iii*) $x = h (\cot \theta_1 - \cot \theta_1)$,

whence

$$h = \frac{x}{\cot \theta_1 - \cot \theta_2} \qquad ...(v)$$

Knowing x, θ_1 and θ_2, *h* can be calculated. It is assumed that the base line is perpendicular to the object.

Measurement of the Altitude of the Sun: To find the altitude of the sun at sea, the horizon being definite is taken as one object, the sun itself being the other. On land the horizon line is not definite so an *artificial horizon is* formed by a pool of mercury contained in a flat vessel. Since it is highly reflective and being heavy it is not liable to be easily disturbed by the wind. The angle subtended at the eye of the observer by the sun and its image in the mercury is measured. It is obviously equal to twice the altitude of the sun.

Microscopes, Telescopes and Eyepieces: *The Visual Angle and Visual Acuity:* The apparent size of any object as seen with the unaided eye depends on the angle it subtends at the eye. It is called the visual angle. The retinal image becomes larger and larger as the object is brought closer to the eye on account of the increase in this visual angle.

Three positions of the same object are shown and we find that the size of the retinal image increases as the object approaches the eye,

$$O'C > O'B' > O'A'.$$

But there is a limit as to how close an object can be brought to the eye if the latter is still to produce a sharp image without over strain. For a normal eye this distance of most distinct vision is taken as 25 cm, and the limiting angle which two point objects at this distance must subtend at the eye in order that they may be distinguished as separate is between one to two minutes and is called the visual acuity or the resolving power of the eye.

The purpose of the magnifying instruments like microscopes and telescopes is to increase the range of the eye by forming a virtual image of the object which subtends a much greater visual angle. The eye sees the image instead of the object directly.

The magnifying power of an optical instrument measures the ratio of the visual angle of the image as seen through the instrument to that of the object as seen by the naked eye.

Magnifying Glass: A simple convex lens of short focus is sometimes used as a magnifier. The object *AB*, to be examined is placed close, to the eye and the short focus convex lens is interposed so that the object is within the principal focus of the lens and at a distance *u* from it. Its virtual image is formed at the distance of distinct vision (*D*). Assuming the eye very close to the lens, the image will also be at a distance *D* from the lens. Let α be the angle which the object *AB* subtends when at this distance and β the angle subtended by the image *A′B′* at the eye.

Then for the lens *E*, $\frac{1}{v} - \frac{1}{u} = \frac{1}{f}$, where *f* is its focal length.

Since *conventionally*, *v* and *u* are both negative, and $v = D$,

$$-\frac{1}{d} + \frac{1}{u} = \frac{1}{f},$$

$$\frac{D}{u} = 1 + \frac{D}{f} \qquad ..(a)$$

By definition, the magnifying power is given by

$$M = \frac{\text{angle subtended by image}}{\text{angle subtended by the object placed at the distance of distinct vision}}$$

$$= \beta / \alpha.$$

Now, $$\beta = \frac{h'}{D} \text{ and } \alpha = \frac{h}{D},$$

$$\therefore \quad M = \frac{\beta}{\alpha} = \frac{h'}{h} = \frac{D}{u}, \quad \ldots$$

[by similar Δs $A'EB'$ and AEB]

$$v = D$$

whence by equation (*a*) above $M = 1 + \dfrac{D}{f}$ ...(2)

$$v = D$$

With the image formed at the distance of distinct vision, the eye has to exert its maximum accommodative power. For perfectly relaxed normal eye the far point is at infinity so that the object must be placed at the focus of the lens. In this case,

$$\beta = \frac{h}{f} \text{ and as before, } \alpha = \frac{h}{D}$$

$$\therefore \quad M = \frac{\beta}{\alpha} = \frac{D}{f}. \quad \ldots (3)$$

$$v \to \infty$$

The magnifying power is, therefore, larger if the image is formed at the distance of distinct vision. But magnifying glasses have generally short focal lengths so there is not much difference between the two expressions and the simpler expression $\dfrac{D}{f}$ $= \dfrac{25}{f}$ is commonly used in labelling the power of the magnifiers.

Astronomical Telescope: A refracting type astronomical telescope consists of a convergent object glass of long focal length and a convex eyepiece of short focal length, mounted on the same axis at the opposite ends of a metallic tube. The eyepiece can be slided in or out as desired.

Rays from a distant object are brought to focus in the focal plane of the object glass forming a real image. The eyepiece is used to magnify this real image and to form a virtual image at any distance between 25 cm and infinity.

We have already noted that for a normal-sighted eye objects are viewed most easily by parallel rays, so we shall assume that rays become parallel after emerging from the eyepiece. The telescope is then said to be in normal adjustment.

The optical path of two bundles of rays through the telescope, in this case the rays *AAA* coming from one part and *BBB* from the other part form the inverted real image *ba* in the second focal plane of the object glass. If this is also the first focal plane of the eye lens, these rays spread out from *ab* and eventually on emergence from the eye lens come out as two bundles of parallel rays. When these emergent bundles fall on the eye, the final inverted and magnified image is seen at *A′B′*.

The magnifying power of the telescope is defined as the *ratio of the angle subtended at the eye by the final image A′B′ to the angle subtend at the eye by the object itself.* Let the object subtend an angle α at the objective then it would subtend approximately the same angle at the eye, as the distance between the eye and the objective is negligible compared to the distance of the object from the telescope. Let the angle subtended at the eye by the final image be β.

By definition, magnifying power. (M) = β / α.

Now $\beta = \angle \beta Ea = \frac{ab}{f}$, where f is the focal length of the eyepiece and $\alpha = \angle AOB = \angle bOa = \frac{ab}{F}$, where F is the focal length of the objective.

$$\therefore \qquad M = \frac{\beta}{\alpha} = \frac{ab}{f} \div \frac{ab}{F} = \frac{F}{f}. \qquad \text{... (4)}$$

Hence, the *magnifying power of a telescope in normal adjustment is the ratio between the focal length of the objective and that of the eyepiece.*

When a telescope is employed to see near objects as in our laboratory experiments, then it is not in normal adjustment and the final image may be caused to be at any desired position convenient to the observer.

E represent the objective and the eye lens of a telescope respectively, *AB* is the object at a distance *u* from the objective, *ba* its image in the objective and $B_1 A_1$ the final image as seen through the system at a distance *v*, say, from the eye lens. The magnifying power is given by

$$m = \frac{\beta}{\alpha} = \frac{\text{angle subtended by the image at the eye}}{\text{angle subtended by the object seen directly by the eye}}$$

If the eye is placed close to the eye lens then assuming that the length of the telescope tube is small as compared to *u*, the angle subtended by the image at the eye $(\beta) = \frac{h}{u_1}$ by the object $(\alpha) = \frac{h}{x}$.

$$\therefore \qquad M = \frac{\beta}{\alpha} = \frac{h}{u_1} \div \frac{h}{x} = \frac{x}{u_1},$$

where *x* is obtained by the relation (using numerical values),

$$\frac{1}{x} + \frac{1}{u} = \frac{1}{f} \text{ for the objective,}$$

$$\text{whence} \qquad x = \frac{uF}{u - F}.$$

u_1 is obtained by the relation, $-\frac{1}{v} + \frac{1}{u_1} = \frac{1}{f}$ for the eye lens, whence $$u_1 = \frac{vf}{u+f}.$$

Substituting these values of x and u_1 in the above equation, we have

$$M = \frac{uF}{u-F} \times \frac{v+f}{vf}. \qquad ...(5)$$

Special Cases

1. When the object is at a distance u and the image is formed at the least distance of distinct vision ($v = D$), equation (5) may be written as

$$M = \frac{uF}{u-F} \times \frac{D+f}{Df}.$$

2. When the object is at a distance u and the image is at infinity ($v \to \infty$), equation (5) may be written as

$$M = \frac{uF}{u-F} \cdot \frac{1+\frac{f}{v}}{f}.$$

or $$M_{v\to\infty} = \frac{u\,F}{u-F} \times \frac{1}{f} = \frac{u}{u-F} \cdot \frac{F}{f}.$$

Since $\frac{u}{u-F}$ is greater than 1, the magnifying power for objects at finite distances is greater than F/f, the magnifying power in *normal* adjustment.

3. When the object is at infinity ($u \to \infty$) and the image is at the least distance of distinct vision $v = D$, equation (5) may be written as

$$M = \frac{F}{1-\frac{F}{u}} \times \frac{D+f}{D}$$

or
$$\underset{u \to \infty}{M} = \frac{F}{f} \cdot \frac{D+f}{D}.$$

4. When both u and v are infinite, the telescope is in normal adjustment, then equation (5) gives

$$M = \frac{F}{1 - \frac{F}{u}} \times \frac{1 + \frac{f}{v}}{f}$$

$$= \frac{F}{f}, \text{ a value obtained already.}$$

The Exit Pupil and the Entrance Pupil: We find that all rays which pass through both the lenses, pass through the area, which is in fact the image of the apertura of the object glass in the eye lens. This area is called the exit pupil or the eye ring or the Ramsden's circle and the point where the area intersects the axis of the telescope is called the eye point.

In telescopes object glass is the aperture stop and is, therefore, the entrance pupil. The eye ring is in fact the image of the object glass and all rays which enter the object glass must emerge through the eye ring. The exit pupil is, therefore, the place where the eye must be situated to see the largest range of objects. If eye is placed on either side of the eye point, the outer bundle of rays would not fall on the eye at all and so the outer parts of the object would not be visible. The instrument is generally designed with its eye ring smaller than the eye pupil so that all the light from the object glass enters the eye. It is important, therefore, to know the diameter KJ of the eye ring. Let it be d and that of the object glass be D.

Now, $KJ = LN\ \theta \times f$, where $\theta = \angle\ LbN$ and f is the focal length of the eyepiece. Also $\theta = D/F$, where F is the focal length of the objective.

$$\therefore \quad KJ = d = \frac{D}{F} \times f = \frac{D}{F/f} = \frac{D}{M}. \qquad \dots (9)$$

That is, the diameter of the eye ring varies *directly* as the diameter of the objective and *inversely* as the magnifying power.

We shall see later the resolving power of optical instruments that for good definition of the image and large resolving power of the instrument the use of objectives with large apertures is a great advantage. Consequently, increase in the diameter of the eye ring may be limited by increasing the magnifying power of the telescope. But again the magnifying power should not exceed that at which the size of the exit pupil matches that of the eye. This particular value is called the normal magnifying power.

It is at this value that the illumination is maximum for the field of view is seen by full pencils and it is the largest.

The diameter of pupil of the eye is generally taken to be 2.54 mm so that the size of the exit pupil should be less than 2-54 mm but it should not be less than 085 mm, for below it the image will not be bright enough.

Then, $\dfrac{Df}{F} \ngtr 2.54$ mm; if $D = 12.70$ cm, $\dfrac{f}{F} \ngtr \dfrac{1}{50}$ or $\dfrac{F}{f} \nless 50$.

If $\dfrac{Df}{F}$ is still smaller, $\dfrac{F}{f}$ can be made still larger.

In Yerkęs' refracting telescope the diameter D of the aperture of the objective is 101.6 cm (40 inches)

so that if $\dfrac{Df}{F} \ngtr 2.54$ mm, $\dfrac{F}{f} \nless 400$.

Also if $\dfrac{Df}{F} \nless 0.85$ mm, then $\dfrac{F}{f} \ngtr 1200$,

so that the minimum and maximum magnifying powers are 400 and 1200 respectively.

From this discussion, it is obvious that if F, f and D are subject Df mainly to the restriction that $\dfrac{Df}{f} \ngtr 2.54$ mm, that is, there is no upper limit to the magnifying power of the telescope. We find, however, that beyond a certain point nothing is gained by increasing magnification, the image becomes larger but reveals no fresh details. The focal length of the objective and its aperture both should be increased. We shall come to this subject again in the study of the resolving power of the optical instruments.

Example: *A spectrometer telescope objective of diameter 3.5 cm and focal length 35 cm is fitted with an eyepiece of focal length 4 cm. Find the magnifying power of the telescope and the diameter of Ramsden's circle.*

Here F = 35 cm, f = 4 cm, D = 3.5 cm.

The magnifying power, $M = \frac{F}{f} = \frac{35}{4} = 8.75.$

The diameter of Ramsden's circle, $d = \frac{D}{F/f} = \frac{D}{M}$

$$= \frac{3.5}{8.75} = 0.4 \text{ cm.}$$

Reflecting Telescope: Most of the large astronomical telescopes of the world today employ concave paraboloidal mirrors in place of achromatic glass objectives. They have several advantages over the refracting type.

1. A concave reflector is free from chromatic aberration and also spherical aberration if paraboloidal in form, thus permitting a very high relative aperture and so giving a very high resolving power. This is a point of great importance in the study of stars.
2. Large aperture lenses of pure optical glass of sufficient homogeneity are difficult to be manufactured. Furthermore, an achromatic objective must be constructed of at least two lenses which would require the grinding and polishing of four surfaces while for a mirror only a single surface requires finishing. The objective of the

world's largest refracting telescope at Yerkes is only 101.6 cm (= 40″) in diameter.

3. Greater stability of the telescope is obtained by having the largest and the heaviest optical part at the bottom. The movements are electrically controlled and this has made the observations easier and more accurate.

The object glass is a concave mirror C which reflects the parallel rays coming from the distant object to a small plane mirror M at an inclination of 45° to the axis so that the real image is formed in its front at $a'b'$. It is viewed through the eyepiece E whose axis is at right angles to the axis of the instrument and the final image is formed at $A'B'$. The magnifying power is the ratio of the angles $a'Eb'$ and bCa, taking the latter as the angle which the object would subtend at the eye if viewed directly.

$$\therefore \qquad M = \frac{\angle a'Eb'}{\angle aCb} = \frac{a'b'}{f} \div \frac{ab}{F} = \frac{F}{f}, \qquad \ldots [\because ab = a'b']$$

where F and f are the focal lengths of the concave mirror and the eyepiece respectively.

Other reflecting telescopes based on the same principle have been designed by Herschel, Cassegrain and Gregory. In Herschel's telescope the concave reflector is placed obliquely to the tube and the image is formed in its focal plane near the edge of the tube and viewed directly through an eyepiece. In Cassegrain's type, the light reflected from the concave mirror converges and meets a small convex mirror m before reaching the focus at F. The focal length of the convex mirror is such that a real image is formed just outside a hole at the vertex of *the* concave mirror, where it is viewed by an eyepiece.

Such an arrangement considerably increases the effective focal length of the objective and so gives a greater magnifying power to the telescope. The aperture of the convex mirror is of such dimensions that it just receives all the rays reflected from the concave mirror to F. The diameter of the hole at the centre of the concave mirror should not be greater than the aperture of the convex mirror for otherwise some of the incident light would enter directly into the eyepiece.

Gregory's type is similar to Cassegrain's but the small mirror is concave instead of convex to the light so that its vertex must be outside the focal length of the large concave mirror.

The Mount Palomer Telescope: The optical features of the great 5.08 m. aperture reflecting telescope. It is of the Cassegrain's type. With the large concave mirror C parallel rays entering the telescope would be brought to focus at F. But before reaching F the converging bundle of rays is reflected by a convex mirror m down the axis and passing through a hole in the objective comes to focus at S, a point just outside the hole so that F and S are conjugate foci for the convex mirror m. For visual observations the image S falls in the first focal plane of the eyepiece, so that the final image is at infinity. This image is inverted since the second image at S is erect with respect to the first image at F and this first image is inverted with respect to the object.

If the image is to be photographed, at the point S is located a photographic plate or the slit of a spectrograph. The mirror has the light-gathering power of a million eyes and it may be regarded as a giant camera which has enabled astronomers to discover objects a thousand million light years away from the earth. This method has the advantage that the plates can be exposed for hours, if necessary, to the object to be studied and thus enabling permanent records to be made.

It excels in performance the 2.54 m Mount Wilson reflecting telescope and has opened up a view beyond our own stellar system towards hazy patches very much like our own milky way. They are known as extra Galactic nebulae—independent stellar systems some where in the infinite Universe. Within the orbit of practical research thus came all aspects of stellar energy, building up and breaking down of elements in space, the problem of abundance of various chemical elements in stars and the investigation of the theories of the Universe.

The telescope has a focal length of 72.24 m and is located in the dome of California observatory which is 41.76 m in diameter and 41.15 m high. The revolving part alone weighs 1000 tons but is easily controlled and moved by comparatively diminutive electric motors.

Formulae and Method

Two Thin Lenses Separated by a Finite Distance

In the theory of optical instruments it is often essential to make use of two or more lenses separated by a distance for the removal of certain aberrations in the image formed by the lens system. We shall, therefore, consider the image formation by such complex systems confining our attention only to paraxial rays and approach this subject by studying the behaviour of two thin lenses having a common axis and separated by some finite distance.

It is possible to replace the combination by a single thin lens which, when placed at a suitable fixed point, produces an image of *the same size but not generally in the same position* as that produced by the combination. It is in this *restricted* sense, namely, that it produces the same transverse magnification as does the combination tor all small objects normal to its axis that the lens is said to be an equivalent lens and its focal length is said to be the equivalent focal length.

Equivalent Focal Length of a Convex Lens System: Let L_1, L_2, be two thin convex lenses of focal lengths f_1, f_2 placed on a common axis a distance 'a' apart.

Consider a ray *PA* incident on L_1, parallel to the axis at a height $O_1A = h_1$, above the axis. It suffers a deviation through an angle δ_1 at the first lens, and proceeds towards its second principal focus F_{sa} meeting the second lens at a height $O_2B = h_2$, where it is further deviated through an angle δ_2. It then intersects the axis in F_2 the second principal focus of the lens system.

Let *PA* produced, cut F_2B in M_2, then a single convex lens placed in the position M_2H_2 and having focal length H_2F_2 is equivalent to the lens system. Thus, H_2F_2 (= *F*) is the required equivalent focal length, we have

$$\text{deviation by first lens, } \delta_1 = \frac{h_1}{f_1},$$

$$\text{deviation by second lens, } \delta_2 = \frac{h_2}{f_2},$$

$$\text{anddeviation by equivalent lens, } \delta = \frac{h_1}{F}.$$

Also $$\delta = \delta_1 + \delta_2.$$

$$\therefore \quad \frac{h_1}{F} = \frac{h_1}{f_1} + \frac{h_2}{f_2}. \quad \text{... (i)}$$

Now, $$h_2 = O_2B = O_2K - BK$$

$$= h_1 - a\delta_1 = h_1 - a\frac{h_1}{f_1}$$

$$= h_1\left(1 - \frac{a}{f_1}\right).$$

Putting the value of h_2 in (*i*), we have

$$\frac{h_1}{F} = \frac{h_1}{f_1} + \frac{h_1}{f_2}\left(1 - \frac{a}{f_1}\right).$$

or $$\frac{1}{F} = \frac{1}{f_1} + \frac{1}{f_2} - \frac{a}{f_1 f_2} \quad \text{... (1)}$$

$$= \frac{f_1 + f_2 - a}{f_1 f_2} \qquad \text{... (1a)}$$

whence
$$F = +\frac{f_1 f_2}{f_1 + f_2 - a} \qquad \text{... (2)}$$

or
$$F = \frac{-f_1 f_2}{\Delta},$$

where $\Delta = a - (f_1 + f_2)$ and is known as the Optical separation or optical interval between the two lenses.

Let us find the position of the equivalent lens, *i.e.*, distance O_2H_2 or O_1H_3.

The Δs $BF_2\ O_2$ and $M_2F_2H_2$ are obviously similar.

$$\therefore \qquad \frac{h_2}{h_1} = \frac{O_2F_2}{H_2F_2}$$

or
$$O_2F_2 = \frac{h_2}{h_1}.$$

$$H_2F_2 = \left(1 - \frac{a}{f_1}\right)F. \quad \ldots \left[\because \frac{h_2}{h_1} = 1 - \frac{a}{f_1}\right].$$

Now,
$$O_2H_2 = H_2F_2 - O_2F_2$$

$$= F - F\left(1 - \frac{a}{f_1}\right) = F\frac{a}{f_1}$$

$$= \frac{f_1 f_2}{f_1 + f_2 - a} \times \frac{a}{f_1} = \frac{af_2}{f_1 + f_2 - a}.$$

Let us conventionally put it equal to $-\beta$ as O_2H_2 is to the left of lens O_2.

$$\therefore \qquad -\beta = \frac{af_2}{f_1 + f_2 - a},$$

or $$\beta = -\frac{af_2}{f_1 + f_2 - a} = -\frac{aF}{f_1} \quad \text{... (3)}$$

and $$O_1H_2 = O_1O_2 - O_2H_2$$

$$= a - \frac{Fa}{f_1} = a\left(1 - \frac{F}{f_1}\right)$$

Thus, the equivalent lens must be placed at a distance $-\frac{aF}{f_1}$ in front of the second lens or a distance $a\left(1 - \frac{F}{f_1}\right)$ behind the first lens.

The plane M_3H_2 where the deviation of the incident parallel ray coming from the left appears to occur is called the second principal plane of the lens system and the point H_2, where it cuts the axis, the second principal point. *The second principal focus F_2 is situated at a distance P towards the right of this point.*

In the same way by considering a ray parallel to the axis incident from the right-hand side we could find the position of F_1 the point where it intersects the principal axis after emerging from the lens system. It is called the first principal focus of the lens system. In the same way as above the position of M_1H_1, the first principal plane and hence the point H_1, the first principal point of the lens system could be determined.

As above we will find that

$$F = +\frac{f_1 f_2}{f_1 + f_2 - a}$$

$$O_1H_1 = \frac{Fa}{f_2} = \frac{f_1 f_2}{f_1 + f_2 - a} \times \frac{a}{f_2} = \frac{af_1}{f_1 + f_2 - a}$$

Let this distance be denoted by α, then

$$\alpha = \frac{af_1}{f_1 + f_2 - a} = +\frac{af}{f_2} \quad \text{...(4)}$$

and $$O_1H_1 = a\left(1 - \frac{F}{f_2}\right).$$

We thus find that the equivalent focal length is independent of the direction from which the light enters the lens system. It also becomes evident that the behaviour of the system cannot be reproduced in all respects by an equivalent thin lens in a fixed position, for the equivalent lens would need to be placed in the plane M_2H_2 to deal with a ray parallel to the axis from the left while it must be placed in the plane M_1H_1 to deal with a ray parallel to the axis from the right in the way as the system does. *The first principal focus F_1 is situated at a distance F towards the left of this point H_1.*

(a) *Power of the Coaxial Systems:* If P_1 and P_2 are the powers of the component lenses and P the power of the combination, then by equation (1)

$$P = P_1 + P_2 - aP_1P_2 \qquad ...(5)$$

Equivalent Focal Length of a Concave Lens System: Let, as before, L_1, L_2 be two thin concave lenses placed on a common axis at a distance 'a' apart.

Consider a ray PA incident on L_1 parallel to the axis at a height $O_1A = h_1$ above the axis. It suffers a deviation δ_1 at the first lens and on diverging appears to proceed from its second principal focus F_{2a}. Meeting the second lens at a height $O_2B = h_2$, it is further diverged through δ_2 and appears to intersect the axis in F_z, the second principal focus of the lens system.

Let PA produced cut F_2B in M_2, then a single concave lens placed in the position M_2H_2 and having focal length H_2F_2 is equivalent to the lens system. Thus, H_2F_2 (=F) is the required equivalent focal length, we have

$$\text{deviation by first lens, } \delta_1 = \frac{h_1}{f_1}$$

$$\text{deviation by second lens, } \delta_2 = \frac{h_2}{f_2}$$

and deviation by equivalent lens, $\delta = -\dfrac{h_1}{F}$.

Also $\qquad \delta = \delta_1 + \delta_2,$

$$\therefore \qquad -\frac{h_1}{F} = -\frac{h_1}{f_1} - \frac{h_2}{f_2},$$

or
$$\frac{h_1}{F} = \frac{h_1}{f_1} + \frac{h_2}{f_2} \qquad \text{... (1)}$$

Now,
$$h_2 = O_2B = O_2K + KB$$
$$= h_1 + a\delta_1$$
$$= h_1 - a\frac{h_1}{f_1} = h_1\left(1 - \frac{a}{f_1}\right).$$

Putting the value of h_2 in (*i*), we have

$$\frac{h_1}{F} = \frac{h_1}{f_1} + \frac{h_1}{f_2}\left(1 - \frac{a}{f_1}\right)$$

or
$$\frac{1}{F} = \frac{1}{f_1} + \frac{1}{f_2} - \frac{a}{f_1 f_2},$$

whence
$$F = +\frac{f_1 f_2}{f_1 + f_2 - a}.$$

If P_1 and P_2 be the powers of the component lenses and P the power of the combination, then

$$P = P_1 + P_2 - aP_1 P_2.$$

These relations are the same as expressed in equations (1), (2) and (5) in the case of the convex lens system.

The determination of positions of the principal planes and principal points in this case is left as an exercise for the students.

All the formulae derived above in respect of convex and concave systems also hold true when the component lenses are of opposite sign.

Relation between the positions of conjugate points and equivalent focal length in a coaxial lens system separated by a finite distance. The usual thin lens formula

$$\frac{1}{V}-\frac{1}{U}=\frac{1}{F}.$$

also holds in the case of coaxial lens system consisting of two or more lenses not in contact provided, however, U and V are measured from first and second principal points H_1 and H_2, respectively, *viz.*, $U = u - \alpha$, and $V = v - \beta$, where α is the distance of the first principal plane from the first lens and β is the distance of the second principal plane from the second lens, u and v are the distances of the object from the first lens and that of image from the second lens , respectively. The distances U and V are called the reduced object and image distances , respectively.

Let f_1 and f_2 be the focal lengths of the two component lenses separated by a distance a. Consider an object u cm away from the first lens. Then, if v' be the distance of the image formed by this lens and v is the distance of the final image from the second lens, we can write (remembering that the image due to the first lens is the object for the second),

$$\frac{1}{v'}-\frac{1}{u}=\frac{1}{f_1}$$

and $$\frac{1}{v}-\frac{1}{v'-a}=\frac{1}{f_2}.$$

Eliminating v' from these equations, we have

$$\frac{1}{v-\beta}-\frac{1}{u-\alpha}=\frac{1}{F}, \quad \ldots (6)$$

or $$\frac{1}{V}-\frac{1}{U}=\frac{1}{F} \quad \ldots (7)$$

where $$F=+\frac{f_1 f_2}{f_1+f_2-a},$$

$$\beta = -\frac{f_2 a}{f_1 + f_2 - a},$$

$$\alpha = +\frac{f_1 a}{f_1 + f_2 - a}.$$

The above relations can be deduced by algebra as below :

Now, $$\frac{1}{v'} - \frac{1}{u} = \frac{1}{f_1} \quad \text{... (i)}$$

and $$\frac{1}{v} - \frac{1}{v' - a} = \frac{1}{f_2}. \quad \text{... (ii)}$$

Let us eliminate v'

From (*i*) $$\frac{1}{v'} = \frac{1}{f_1} + \frac{1}{u} = \frac{u + f_1}{u f_1},$$

$$\therefore \quad v' = \frac{u f_1}{u + f_1} \quad \text{... (iii)}$$

From (*ii*) $$\frac{1}{v' - a} = \frac{1}{v} - \frac{1}{f_2} = \frac{f_2 - v}{v f_2},$$

or $$v' - a = \frac{v f_2}{f_2 - v},$$

$$\therefore \quad v' = \frac{v f_2}{f_2 - v} + a \quad \text{... (iv)}$$

Equating the values of v' in (*iii*) and (*iv*)

$$\frac{u f_1}{u + f_1} = \frac{v f_2}{f_2 - v} + a.$$

Multiplying throughout by $(u + f_1)(f_2 - v)$, we have

$$u f_1 (f_2 - v) = v f_2 (u + f_1) + a (u + f_1)(f_2 - v)$$

Opening the brackets and collecting the terms in uv, u and v

$$u f_1 f_2 - u v f_1 = u v f_2 + v f_1 f_2 + u a f_2$$

$$- uva + af_1f_2 - vaf_1,$$

or $$uvf_1 + uvf_2 - uva + uaf_2 - uf_1f_2 - vf_1f_2$$
$$- vaf_1 + af_1f_2 = 0$$

or $$uv\left[f_1 + f_2 - a\right] + u\left[af_2 - f_1f_2\right] + v\left(f_1f_2 - af_1\right) + af_1f_2 = 0$$

Dividing by $(f_1 + f_2 - a)$ throughout, we get

$$uv + \frac{u\left(af_2 - f_1f_2\right)}{f_1 + f_2 - a} + \frac{v\left(f_1f_2 - af_1\right)}{f_1 + f_2 - a} + \frac{af_1f_2}{f_1 + f_2 - a} = 0. \quad \text{... }(v)$$

This equation can be put in the form

$$\frac{1}{v - \beta} - \frac{1}{u - \alpha} = \frac{1}{F}. \quad \text{... }(vi)$$

Multiplying (*vi*) throughout by $F(u - \alpha)(v - \beta)$ and collecting terms in uv, u and v, we obtain

$$uv - (F + \beta)u + (F - \alpha)\, v - F\,(\beta - \alpha) + \alpha\beta = 0. \quad \text{... }(vii)$$

In order that (*v*) and (*vii*) should be identical, we must have

$$- F - \beta = \frac{af_2 - f_1f_2}{f_1 + f_2 - a}, \quad \text{... }(viii)$$

$$F - \alpha = \frac{f_1f_2 - af_1}{f_1 + f_2 - a} \quad \text{...}(ix)$$

and $$-F\beta + F\alpha + \alpha\beta = \frac{af_1f_2}{f_1 + f_2 - a}. \quad \text{...}(x)$$

Thus (*viii*), (*ix*) and (*x*) are simultaneous equations to find the three unknowns α, β and F.

To solve these, first multiply the corresponding sides of equations (*viii*), and (*ix*). This gives us

$$-F^2 - F\beta + F\alpha + \alpha\beta = \frac{\left(af_2 - f_1f_2\right)\left(f_1f_2 - af_1\right)}{\left(f_1 + f_2 - a\right)^2} \quad \text{... }(xi)$$

Subtracting (*xi*) from (*x*) gives

$$F^2 = \frac{af_1f_2}{f_1+f_2-a} - \frac{(af_2-f_1f_2)(f_1f_2-af_1)}{(f_1+f_2-a)^2}$$

$$= \frac{af_1f_2(f_1+f_2-a)-(af_2-f_1f_2)(f_1f_2-af_1)}{(f_1+f_2-a)^2}$$

$$= \frac{(f_1f_2)^2}{(f_1+f_2-a)^2}$$

$$\therefore \quad F = \pm\frac{f_1f_2}{f_1+f_2-a} \qquad \text{... }(xii)$$

In order to find whether the positive or negative sign is to be retained with the values of F, it must be noted that equation (*xii*) must be true for all values of a.

When $a = 0$, from (*i*) and (*ii*), we have

$$\frac{1}{v}-\frac{1}{u} = \frac{1}{f_1}+\frac{1}{f_2} = \frac{1}{F},$$

or

$$F = \frac{f_1f_2}{f_1+f_2}.$$

Substituting $a = 0$ in (*xii*) we see that we obtain the value of F just found provided we take the positive sign. Hence,

$$F = +\frac{f_1f_2}{f_1+f_2-a} \qquad \text{... }(xiii)$$

Adding (*viii*) and (*xiii*), we have

$$-\beta = \frac{af_2-f_1f_2}{f_1+f_2-a}+\frac{f_1f_2}{f_1+f_2-a}$$

$$= \frac{af_2}{f_1+f_2-a},$$

or $$\beta = -\frac{af_2}{f_1 + f_2 - a} = -\frac{aF}{f_1}. \quad \text{... }(xiv)$$

Subtracting (*ix*) from (*xiii*), we get

$$\alpha = \frac{f_1 f_2}{f_1 + f_2 - a} - \frac{f_1 f_2 - af_1}{f_1 + f_2 - a}$$

$$= \frac{af_1}{f_1 + f_2 - a} = \frac{aF}{f_2} \quad \text{... }(xv)$$

Hence, also $\alpha f_2 = \beta f_1$ ... (*xvi*)

which is an important relation.

Here α is the distance of first principal plane from first lens and β the distance of second principal plane from second lens.

Formula of Newton

Connecting Conjugate Foci for a Coaxial Lens System: This is an important conjugate-distance relation with reference to principal focal planes.

F_1, F_2 are the principal foci and H_1, H_2 the unit points. P_1Q_1 is the image of an object PQ, perpendicular to the axis drawn. Let $PQ = h_1$ $P_1Q_1 = h_2$, $F_1Q = x_1$ and $F_2Q_1 = x_2$ and let ϕ_1 and ϕ_2 be the first and second focal lengths of the system. According to our convention h_1, ϕ_2, x_2 are positive and h_2, ϕ_1, x_1 are negative.

Now, Δs PQF_1 and $B_1H_1F_1$ are similar.

$$\therefore \quad \frac{H_1B_1}{PQ} = \frac{H_1F_1}{F_1Q},$$

or $$\frac{-h_2}{h_1} = \frac{-\phi_1}{-x_1} = \frac{\phi_1}{x_1}, \quad \text{... }(i)$$

Also, Δs $A_2H_2F_2$ and $P_1Q_1F_2$ are similar.

$$\therefore \quad \frac{P_1Q_1}{H_2A_2} = \frac{F_2Q_1}{H_2F_2},$$

or $$\frac{h_2}{h_1} = \frac{x_2}{\phi_2}. \qquad ...(ii)$$

Comparing equations (*i*) and (*ii*)

$$\frac{\phi_1}{x_1} = \frac{x_2}{\phi_2},$$

or $$x_1 x_2 = \phi_1 \phi_2.$$

When the medium is the same on both sides

$$\phi_1 = -\phi_2,$$

$\therefore$ $$x_1 x_2 = \phi_2^2 = -F^2 \qquad ...(19)$$

since F is being used for f_2 in the text.

Also from (*i*) and (*ii*)

$$m = \frac{h_2}{h_1} = \frac{+F}{x_1} = \frac{-x}{F} \qquad ...(20)$$

Example: *Two convex lenses with focal lengths 10 cm and 20 cm ore placed co-axially 5 cm apart. Find (a) the equivalent focal length of combination, (b) the positions of the principal planes, focal planes, and (c) the position of the image of an object placed 60 cm from the first lens.*

(*a*) Here $f_1 = +10$ cm, $f_2 = +20$ cm and $a = 5$ cm.

$\therefore$ $$H_2F_2 = F = +\frac{f_1 f_2}{f_1 + f_2 - a}$$

$$= +\frac{10 \times 20}{10 + 20 - 5} = +\frac{200}{25} = +8 \text{ cm.}$$

and $$H_1F_1 = -8 \text{ cm.}$$

$$O_2H_2 = \beta = -\frac{af_2}{f_1 + f_2 - a}$$

$$= -\frac{5 \times 20}{10 + 20 - 5} = -\frac{100}{25} = -4 \text{ cm.}$$

$$O_1H_1 = \alpha = +\frac{af_1}{f_1 + f_2 - a}$$

$$= +\frac{5 \times 10}{10 + 20 - 5} = +\frac{50}{25} = + 2 \text{ cm.}$$

(*b*) Using these values the positions of the principal points and focal points have been indicated

(*c*) Since $u = - 60$ cm, we can now write

$$U = u - \alpha = - 60 - 2 = - 62 \text{ cm},$$
$$F = + 8 \text{ cm}, V = ?$$

Applying the relation,

$$\frac{1}{V} - \frac{1}{U} = \frac{1}{F}, \text{ we have } \frac{1}{V} - \frac{1}{-62} = \frac{1}{8}$$

$$\therefore \qquad \frac{1}{V} = \frac{1}{8} - \frac{1}{62},$$

whence $$V = \frac{8 \times 62}{54} = 9.2 \text{ cm.}$$

$$v - \beta = 9.2 \text{ cm}$$

or $$v - (-4) = 9.2 \text{ cm.}$$

or $$v = 5.2 \text{ cm.}$$

That is, the final image is 5.2 cm beyond the second lens.

$$m = \frac{V}{U} = \frac{9.2}{-62} = - 0.15.$$

Alternatively: The same result may be arrived at from *first* principles, as below:

For the first lens

$$\frac{1}{v_1} - \frac{1}{n_1} = \frac{1}{f_1},$$

where $u_1 = - 60$ cm; $f_1 = + 10$ cm.

$$\therefore \qquad \frac{1}{v_1} - \frac{1}{-60} = \frac{1}{10}$$

or $$\frac{1}{v_1} = \frac{1}{10} - \frac{1}{60} = \frac{6-1}{60} = +\frac{5}{60}$$

or $$v_1 = +12 \text{ cm.}$$

This image serves as a virtual object for the second lens at a distance (+12 – 5) = + 7 cm from it.

For refraction through the second lens

$$u_2 = +7 \text{ cm}; f_2 = +20 \text{ cm.}$$

Now, $$\frac{1}{v_2} - \frac{1}{u_2} = \frac{1}{f_2},$$

$$\therefore \quad \frac{1}{v_2} = \frac{1}{v_2} - \frac{1}{+7} = \frac{1}{20},$$

or $$\frac{1}{v_2} = \frac{1}{20} + \frac{1}{7} = \frac{7+20}{20 \times 7} = \frac{27}{20 \times 7}$$

$$\therefore \quad v_2 = \frac{20 \times 7}{27} = 5.2 \text{ cm}$$

whence $$m = \frac{v_1}{u_1} \times \frac{v_2}{u_2} = \frac{12}{-60} \times \frac{5.2}{7} = -0.15.$$

The student should carefully note, that the second principal plane is nearer the first lens than the first principal plane. This is of frequent occurrence and should cause no confusion in the mind of the student, because the object distance is always measured from the first principal plane, wherever it is situated.

Example: *An object is placed at a distance of 50 cm in front of a combination of a thin convex lens of focal length 25 cm and a thin concave lens of focal length 15 cm placed 15 cm apart along the common axis. Find (a) the focal length of the combination, (b) the positions of the cardinal points and (c) the transverse magnification.*

(*a*) Here $f_1 = +25$ cm, $f_2 = -15$ cm and $a = 15$ cm.

$$\therefore \quad H_2F_2 = F = +\frac{f_1 f_2}{f_1 + f_2 - a} = +\frac{25 \times (-15)}{25 - 15 - 15}$$

$$= \frac{375}{-5} = +75 \text{ cm.}$$

and $\quad H_1F_1 = -75$ cm

$$O_2H_2 = \beta = -\frac{af_2}{f_1 + f_2 - a}$$

$$= \frac{15 \times (-15)}{25 - 15 - 15} = -45 \text{ cm.}$$

$$O_1H_1 = \alpha = +\frac{af_2}{f_1 + f_2 + a}$$

$$= +\frac{15 \times 25}{25 - 15 - 15} = -75 \text{ cm.}$$

(*b*) The positions of the cardinal points are marked. Since the lens system is in air, N_1 and N_2 coincide with H_1 and H_2, respectively.

(*c*) Applying the thin lens equation to the convex lens, we get

$$\frac{1}{v_1} - \frac{1}{-50} = \frac{1}{25}$$

or
$$\frac{1}{v_1} = \frac{1}{25} - \frac{1}{50}$$

whence $\quad v_1 = 50.$

This image serves as the virtual object for refraction at the concave lens. The distance of the image from the concave lens is equal to, $u_2 = +50 - 15 = +35$ cm.

Hence, for this lens

$$\frac{1}{v_2} - \frac{1}{35} = -\frac{1}{15}$$

or
$$\frac{1}{v_2} = -\frac{1}{15} + \frac{1}{35},$$

whence $v_2 = -\frac{105}{4}$ cm.

Resultant transverse magnification

$$m = m_1 \times m_2 = \frac{v_1}{u_1} \times \frac{v_2}{u_2}$$

$$= \frac{50}{-50} \times \frac{-105}{4} \times \frac{1}{35} = +\frac{3}{4}.$$

***Example:** Two convex lenses, each, of 20 cm focal length are placed 5 cm apart. A tower 100 metres high and 200 metres distant is viewed through them. Find (a) the focal length of the equivalent lens, and (b) the position and size of image.*

(*a*) The focal length of the equivalent lens is given by

$$F = +\frac{f_1 f_2}{f_1 + f_2 - a} = +\frac{20 \times 20}{20 + 20 - 5}$$

$$= +\frac{400}{35} = +11.43 \text{ cm.}$$

(*b*) To find the position and size of image let us take one lens at a time.

For first lens, $u_1 = -$ 200 metres $= -$ 20,000 cm and $f_1 = +$ 20 cm.

Substituting the values in the relation,

$$\frac{1}{v_1} - \frac{1}{u_1} = \frac{1}{f_1},$$

we have $\frac{1}{v_1} + \frac{1}{20000} = \frac{1}{20},$

or $$\frac{1}{v_1} = \frac{1}{20} - \frac{1}{20000} = \frac{1000 - 1}{20000}$$

$$= +\frac{999}{20000},$$

whence $$v_1 = +\frac{20000}{999} = +\ 20.02 \text{ cm approx.}$$

If u_2 be the distance from the second lens of the image formed by the first,

then $$u_2 = +\ 20.02 - 5.0 = 15.02 \text{ cm.}$$

This serves as an object for the second lens. Substituting this value in the relation

$$\frac{1}{v_2} - \frac{1}{u_2} = \frac{1}{f_2}$$

we get, $$\frac{1}{v_2} - \frac{1}{+15.02} = \frac{1}{20},$$

or $$\frac{1}{v_2} = \frac{1}{20} + \frac{1}{15.02}$$

$$= +\frac{15.02 + 20}{20 \times 15.02} = \frac{35.02}{300.4},$$

whence $$v_2 = \frac{300.4}{35.02} = +\ 8.58 \text{ cm}$$

Now, $$m = \frac{v_1}{u_1} \times \frac{v_2}{u_2} = \frac{20.02}{20000} \times \frac{8.58}{15.02} = \frac{171.77}{300400},$$

i.e., $$\frac{\text{size of image}}{\text{size of object}} = \frac{171.77}{300400}$$

$\therefore$ $$\text{size of image} = \frac{171.77}{300400} \times \text{size of object}$$

$$= \frac{171.77}{300400} \times 10000 = 5.71 \text{ cm.}$$

Note: The reader will come across cases of coaxial systems in which it is not necessary that if the component lenses are converging, the system as a whole needs be converging, it may be both converging or diverging depending upon the positions of the lenses with respect to the incident rays. Thus, the terms 'converging' and 'diverging' used in connection with such systems become *misleading*. Let us illustrate this point by two examples.

Example: *Two thin converging lenses of focal lengths 15 cm and 5 cm are placed 10 cm apart each lens facing the incident ray alternately. Find the positions of the cardinal points and trace the course of a parallel ray through them.*

Taking f_1 = 15 cm, f_2 = 5 cm, and a = 10 cm, we have

$$H_2F_2 = F = \frac{f_1 \times f_2}{f_1 + f_2 - a} = \frac{15 \times 5}{15 + 5 - 10} = 7.5 \text{ cm}$$

and $$H_1F_2 = -7.5 \text{ cm}$$

$$O_1H_1 = \alpha = \frac{af_1}{f_1 + f_2 - a} = \frac{10 \times 5}{15 + 5 - 10} = 15 \text{ cm}$$

$$O_2H_2 = \beta = \frac{af_2}{f_1 + f_2 - a} = -\frac{10 \times 5}{15 + 5 - 10} = -5 \text{ cm.}$$

Thus, if the lenses are arranged as with the cardinal points marked as shown the nodal points coinciding with the principal points let us trace the path of a ray PA_1 parallel to the axis through the system. If PA_1 produced, cuts the first unit plane in B_1, it will proceed along B_2F_2 after emergence from the system where B_2 is on the second principal plane and $H_2B_2 = H_1B_1$ If B_2F_2 cuts the second lens in A_2, the path of the rays between the lenses is A_1A_2 and the emergent ray is the prolongation of B_1A_2. The emergent ray actually crosses the principal axis outside the system which is, therefore, a 'converging' system.

If, now the system is reversed, *i.e.*, the component lenses interchange their positions, the focal points and the principal points occupy the same positions as before relative to the two leases, the words 'first' and 'second' must be interchanged, for with the given data $f_1 = 5$ cm and $f_2 = 15$ cm, $O_1H_1 = \alpha = 5$ cm and $O_2H_2 = \beta = -15$ cm, F remains unaltered. These new positions of the cardinal points are marked In this case also, proceeding as usual we find the course of the emergent ray, and conclude that although the focal length of the system is still positive, the parallel incident ray on emergence does not cross it at a real point, so that the system is a diverging' one.

Example: *An eyepiece is formed by two thin convex lenses of focal lengths 3 cm and 8 cm, and a distance 6 cm apart. Calculate the focal length of the eyepiece to indicate the position of the cardinal points of the eye piece with respect to the lenses.*

We have $f_1 = +3$ cm; $f_2 = +8$ cm and $a = 6$ cm.

$$H_2F_2 = F = +\frac{f_1f_2}{f_1+f_2-a}$$

$$= +\frac{(+3)\times(+8)}{3+8-6}$$

$$= \frac{24}{5} = +4.8 \text{ cm.}$$

and $$H_1F_1 = -4\frac{4}{5} \text{ cm.}$$

$$O_1H_1 = \alpha = +\frac{af_1}{f_1+f_2-a}$$

$$= +\frac{6\times(+3)}{3+8-6}$$

$$= +\frac{18}{5} = +3.6 \text{ cm.}$$

$$O_2H_2 = \beta = -\frac{af_2}{f_1+f_2-a}$$

$$= -\frac{6\times(+8)}{3+8-6}$$

$$= -\frac{48}{5} = -9.6 \text{ cm.}$$

Using these values the positions of principal foci and principal points have been marked. As the lens system is in air the nodal points N_1 and N_2 coincide with the principal points H_1 and H_2, respectively. Hence, the positions of all the six points are marked.

If a ray strikes the first lens at A_1 and the first unit plane in B_1, it must emerge in the direction B_2F_2, where B_2 is on the second unit plane such that $H_2B_2 = H_1B_1$. If B_2F_2 cuts the second lens in A_2, the path of the ray between the lenses is A_1A_2 and the emergent ray is the prolongation of B_2A_2.

Here, F_1 is a real focus and F_2 a virtual focus and the combination is divergent for parallel rays falling *first* on the lens of focal length 3 cm. It can similarly be shown that F_1 is a virtual focus and F_2 a real one and the combination is convergent for rays first falling on the lens of focal length 8 cm.

Minimum Distance between the Object and its Real Image for a Positive Focal Length System of two Lenses: If H_1 and H_2 are the principal points of the system and U and V the object and image distances, as measured from H_1 and H_2, respectively, then x the numerical value of the distance between object and image is given by

$$x = U + V + a, \text{ where } a = H_1H_2.$$

Now, for a positive lens, correcting for signs, we have

$$\frac{1}{V} + \frac{1}{U} = \frac{1}{F},$$

$$\therefore \qquad V = \frac{UF}{U - F}.$$

Hence,

$$x = U + \frac{UF}{U - F} + a \qquad \text{... } (i)$$

For x to be minimum or maximum

$$\frac{dx}{dU} = 1+\frac{F(U-F)-UF}{(U-F)^2}+0 = 0.$$

Solving it, we get $U = 2F$ or $U = 0$.

Differentiating it a second time we will find that $U = 2F$, corresponds to a minimum value of x. The solution $U = 0$ is meaningless as it implies, the contact of the object with the first principal plane. Thus putting the first value of U in (i), we have

$$x_{min} = 2F+\frac{2F^2}{F}+a = 4F + a. \quad \text{... (21)}$$

Example: *A thin converging lens of focal length 15 cm and a thin diverging lens of focal length –10 cm are arranged co-axially at a distance 10 cm apart. Determine the positions of cardinal points.*

Here, $f_1 = 15$ cm, $f_2 = -10$ cm, and $a = 10$ cm

$$H_2F_2 = F = \frac{f_1 f_2}{f_1+f_2-a} = \frac{15\times(-10)}{15-10-10} = +30$$

and $H_1F_1 = -30$ cm.

$$O_1H_1 = \alpha = \frac{af_1}{f_1+f_2-a} = \frac{10\times 15}{15-10-10} = -30 \text{ cm.}$$

$$O_2H_2 = \beta = -\frac{af_2}{f_1+f_2-a} = \frac{10\times(-10)}{15-10-10} = -20 \text{ cm.}$$

The positions of the cardinal points. Since the lens system is in air, the nodal points N_1 and N_2 coincide with the principal points H_1 and H_2, respectively.

Experimental Determination of the Cardinal Points of a Positive Coaxial System in Air: *(a) When the principal foci are real and outside the system.* To locate the principal foci of the lens system O_1 O_2, a plane mirror M and a pin PQ to serve as an object [an illuminated cross wires and a screen may be used instead] are suitably mounted on the optical bench. The lens system is shifted with respect to the pin until there is no parallax between the pin and its inverted image. Thus, the position of $F_{1,}$ the first principal focus of the component lens system is found.

The course of rays for the formation of the inverted image P_1Q_1 of the pin PQ placed in the first focal plane of the lens system. The two will be coincident if the plane of the mirror M is normal to the axis. The interested reader should try to understand the path of the rays through the lens system keeping in view the properties of the cardinal points. A ray PA_1 meeting the first principal plane in B_1 becomes parallel to the axis, cuts the second principal plane in B_t and then proceeds parallel to $PH_{1'}$ intersects the mirror M in E, where it is reflected along EB_2', so that the angle of incidence is equal to the angle of reflection At B_2', a point in the second principal plane, the ray becomes parallel to the axis meeting the first principal plane in B_i'. Join B_1' to $P_{1,}$ a point where the line H_1P_1 drawn parallel to EB_2' cuts PQ produced. Then P_1 is the image of the point P. It can also be proved that the final image P_1Q_1 is *numerically* equal in size to the object PQ, since the Δs PH_1Q and $P_1H_1Q_1$ can easily shown to be congruent.

The position of the second principal focus F_2 is determined in a similar manner by interchanging the positions of the pin and the mirror M. The focal length of the system is next found by any one of the methods given below and the positions of the principal points H_1 and H_2 located, as their distances from the corresponding principal foci are each equal to the focal length. If the system is in air, the positions of the principal points and the nodal points will coincide.

Newton Method of Finding Focal Length: Keeping the system fixed in position and removing the plane mirror, two pins [or an illuminated object and a screen] are arranged one on each side beyond F_1 and F_2 and parallax is removed between one pin and the inverted image of the other. If x_1 and x_2 are the distances of the object and the image , respectively from the corresponding principal foci then $x_1x_2 = -F^2$ whence F may be calculated. The experiment may be repeated for different conjugate points and mean focal length obtained.

Magnification Method: If U and V are the distances of the object and the image from the first and the second principal points , respectively, then

$$m = \frac{V}{U} \text{ and } \frac{1}{V} - \frac{1}{U} = \frac{1}{F}.$$

$$\therefore \qquad 1 - m = \frac{V}{F}. \qquad \text{... } (i)$$

To find the magnification a suitable laboratory method may be employed. If by keeping the lens system fixed we arrange the object and the image scales to give magnifications – 1, –2, –3, etc., then from equation (*i*)

$$1 - (-1) = \frac{V_1}{F} = 2,$$

$$1 - (-2) = \frac{V_2}{F} = 3,$$

$$1 - (-3) = \frac{V_3}{F} = 4,$$

$$\therefore \qquad V_2 - V_1 = V_3 - V_2 = ... = F. \qquad \text{... } (ii)$$

Obviously, therefore, the focal length of the lens system is *numerically* equal to the shift of the image (glass) scale when the magnification is increased by unity. The method is, therefore, applicable when the positions of H_1 and H_2 are unknown and hence U or V cannot be measured directly.

(b) When one of the principal foci is within the system. O_1 and O_2, are the component lenses of the system with F_1 lying in between them and M a plane mirror placed behind them normal to the principal axis. L is an additional lens which produces a real-image of a bright object P on the axis so as to be received by the system which is adjusted until a clear image is formed on the screen at the site of P. The course of rays in this auto-collimating system indicates that the real image formed by L

in the absence of the lens system must occupy the same position as F_1 when the system is in position and arranged as above.

To locate F_2 we proceed as in the first case. Next the focal length of the system is found by the magnification method as Newton's method is not applicable, F_1 being inside, and thus the principal points are located. Since the system is in air, the positions of the principal points and the nodal points are the same.

Method of Nodal Slide

Nodal Slide Method of Determining Cardinal Points of a Thick Lens or a Lens System: The method is based on the special property of the nodal points, *viz., that a ray incident upon the lens system and proceeding towards the first nodal point N_1, gives rise to an emergent ray proceeding from the second nodal point N_2 and parallel to the incident ray.* It leads to the following result that "if a parallel beam of light is incident on a convergent lens system forming an image on a screen at its second principal focus the image does not shift laterally when the system is rotated about a vertical axis passing through its second nodal point."

The principle of the method for a positive lens or a lens system is illustrated where a parallel beam is incident on the lens or a lens system. (*a*) The ray 0 along the axis passes through N_1 and N_2 *to* the focus at Q_2. (*b*) The lens has been rotated through a small angle about an axis through N_2 and the same beam of rays passes through the system coming to a focus at the same point Q_2. The ray 1 is now directed towards N_1 and the ray *O* towards N_2. When projected across from the plane of N_1, to that of N_2 the rays still converge towards Q_2 even though F_2 is now shifted towards one side. Note that ray 1 approaches *N*, in exactly the same direction that it leaves N_2 according to the property stated above.

The special type of device designed for the purpose of making use of this property for finding the nodal points is called a nodal slide.

Nodal Slide: It consists of a heavy metal upright carrying a table which is capable of rotation about the vertical axis of the upright and the angle of rotation can be recorded, if so desired, on its graduated circular box. The lens or lens system is mounted in the holders provided on the table and may slide along a bar graduated in mm on which the position of the lens system is read. This movement is in addition to the fact that the nodal upright can itself side along the optical bench on which it is mounted. The graduated bar can be shifted parallel to itself and clamped in any desired position.

In carrying out the experiment the position of the lens or lens system is adjusted so that the optical axis of the system is made to intersect the axis of rotation of the table. There is a white screen having a narrow cross-slit S on one side of the nodal slide and a plane mirror M on the other side. Monochromatic light from a source is sent through the slit S adjusted to lie at the secondary focal point of the lens system. On emergence as a parallel beam this light is reflected back on itself by the plane mirror M, fixed normally to the optical axis and passes again through the lens system and is brought to focus at a point on the screen adjacent to S. The nodal slide is now rotated about its axis back and forth and the lens system repeatedly moved along the graduated bar until the rotation produces no motion of the image of the slit; the slit screen must be adjusted for each new position of the lens system so that the image is sharply focused upon it. When this condition is reached, the point N_2 where the axis of rotation intersects the principal axis locates one nodal point. The table is next turned through 180° and the process repeated till the image is again focused on the screen which must not be moved and its position remains unaffected by a small rotation of the table. Thus the position of the second nodal point N_1 is found. When the experiment is performed in air the positions of the nodal points are those of the principal points as well and the distance N_2S is the correct focal length.

By this arrangement, therefore, we can locate the nodal points and hence also the principal points. We can also find the focal length accurately and hence the focal points are also located, that is, all the cardinal points are thus known. We can also study by this arrangement the variation in the positions of the nodal points for different separations of the two lenses of a coaxial system.

Principle of Nodal Points

Besides the principal foci and principal points, which are of special importance in developing the theory of coaxial systems of thin lenses, there is another important pair of conjugate points on the axis of such an optical system which may be used to simplify the solution of certain problems. They are so situated that an incident ray directed towards the first will, on emerging from the lens system, proceed (or appear to proceed) from the second point in a direction parallel to that of the incident ray. These points are called the first and second nodal points of the system and are denoted by N_1 and N_2, respectively. They are the points for which the angular magnification is + 1, for the incident ray and its conjugate emergent ray make the same angles with the axis. *When the medium on both sides of the* ***lens*** *system is the same the nodal points and the principal points are coincident.* This can be easily shown by making use of Helmholtz Lagrange's equation according to which

$$\mu_1 h \alpha_1 \equiv \mu_2 h_2 \alpha_2.$$

where the letters have their usual significance. If the medium on both sides is air,

$\mu_1 = \mu_2 = 1$; and since $\alpha_1 = \alpha_2$, we have

$h_1 = h_2$, *i.e.*, the transverse magnification is also unity.

Hence, the nodal points for a combination of lenses placed in air coincide with the principal points. The planes transverse to the principal axis through N_1, and N_2 are called nodal planes- Since nodal points are conjugate, the nodal planes are also conjugate.

The Cardinal Points of a Lens System: In the study of the coaxial systems of lenses we have discussed above some points which have an important bearing on the theory of such systems, namely, the two principal foci, the two principal or unit points and the two nodal points. These six points are called the cardinal points or Gaussian points after the name of a German optician E.F. Gauss, who first investigated the behaviour of thick lenses and coaxial systems of lenses He showed how these points have enabled us to treat optical systems as units and to formulate in their case relationships similar to those for thin lenses or single surfaces giving positions and sizes of the images directly in terms of those of the objects used. By their use we can conveniently avoid the tedious process of treating a coaxial system as a succession of spherical surfaces and considering the image formed by the first surface as the object for the following one and so on, ultimately to determine the size and location of the final image formed by the system.

Besides the cardinal points there are two more pairs of conjugate points of lesser importance with negative linear and angular magnifications. They are respectively the *anti-principal* and *anti-nodal* points.

Graphical Construction of the Image of an Object on the Axis of a Coaxial System: To construct the image of an object in a system of two lenses we may proceed to trace the path of rays through each lens separately. Such a method is rather tedious, so in practice the actual positions of the lenses are abandoned and we make use of the properties of the principal foci and principal planes which are planes of unit magnification, *viz.*, the emergent ray proceeds or appears to proceed from a point in the second principal plane at the same distance from the principal axis of the system and on the same side of it as is the point in the first principal plane to which the incident ray proceeds or appears to proceed. Knowing therefore the positions of the principal foci and unit points, we can easily construct the image of a small object on the axis of a coaxial system.

Let us consider in turn a system of two coaxial lenses in which the focal length is (a) positive, and (*b*) negative.

When the Combination is Positive: Let PQ denote a small object at right angles to the principal axis at Q, the cardinal points F_1, F_2, H_1 and H_2 of the system being as indicated. Then a ray PA_1 parallel to the axis meeting the first principal plane in A_1 will emerge through a point A_2 in the second principal plane such that $H_2A_2 = H_1A_1$ and will pass through F_2, the second principal focus of the lens system. Also a ray PB_1 passing through F_1 and meeting the first principal plane in B_1 will emerge from the system in the direction B_2P_1 parallel to the axis so that $H_2B_2 = H_1B_1$. These two rays intersect in P_1, the image of P, and P_1Q_1 drawn perpendicular, to the axis is the image of PQ. The image is real but inverted.

When the combination is negative. Let PQ as before be a small object perpendicular to the axis of the lens system. The paths of two rays from P, one PA_1 parallel to the axis and the other PB_1 directed towards F_1 are traced in the usual way so that the first ray emerges out of the system at a point A_2 and appearing to proceed from F_2 and the second ray on emergence becomes parallel to the axis appearing to intersect the first ray at P_1. P_1Q_1 is the image of PQ. It is virtual, erect and small in size.

It may also be noted that an object ray directed towards H_1 has a corresponding image ray whose direction passes through H_2 and is parallel to that of the object ray.

Images in the case of coaxial lens systems are not generally in the same position as those produced by the equivalent lens. Let us consider in (*a*) the image of an object PQ is formed by a two-lens system of positive focal length by the method explained above and in *(b)* the image of the same object is formed by a single lens of the same focal length as the combination.

Conjugate Relations Referred to the Principal Planes of a Coaxial System: Consider a coaxial system with media of refractive indices μ_1 and μ_2 on its either side. *Let* F_1 and F_2 be the principal foci and H_1 and H_2, the principal points of a convergent coaxial system. Let PQ be an object perpendicular to the principal axis at Q. To obtain the image of PQ let us consider two rays:

(1) A ray PA_1, parallel to the axis meeting the first principal plane in A_1 will emerge through the point A_2 in the second principal plane so that $H_2A_2 = H_1A_1$ and will pass through F_2, the second principal focus of the lens system.

(2) The ray PB_1, passing through F_1 and meeting the first principal plane in B_1, will emerge in the direction B_2P_1 parallel to the axis so that $H_2B_2 = H_1B_1$.

These rays intersect in P_1, the image of P, and P_1Q_1 drawn perpendicular to the axis is the real image of PQ. Let U and V denote , respectively the distances of the object and the image from the corresponding principal points. According to our convention the signs to be attached to the various distances are as follows:

$$H_1Q = -U,\ H_2Q_1 = V; PQ = n_1,\ P_1Q_1 = -h_2,$$

$$H_1F_1 = -\phi_1,\ H_2F_2 = +\phi_2,$$

where ϕ_1 and ϕ_2 are the first and second principal focal lengths of the lens system.

Now, Δs $F_1H_1B_1$ and PA_1B_1 are similar.

$$\therefore \quad \frac{H_1F_1}{A_1P} = \frac{H_1B_1}{A_1B_1} = \frac{H_1B_1}{H_1A_1 + H_1B_1}$$

or $$\frac{-\phi_1}{-U} = \frac{-h_2}{h_1 - h_2}. \quad \text{... } (a)$$

Also, Δs $H_2A_2F_2$ and $B_2A_2P_1$ are similar.

$$\therefore \quad \frac{H_2F_2}{B_2P_1} = \frac{H_2B_2}{B_2A_2} = \frac{H_2A_2}{H_2A_2 + H_2B_2},$$

or $$\frac{+\phi_2}{V} = \frac{h_1}{h_1 - h_2}. \quad \text{... } (b)$$

Adding (*a*) and (*b*)

$$\frac{-\phi_1}{-U} + \frac{\phi_2}{V} = \frac{h_1}{h_1 - h_2} - \frac{h_2}{h_1 - h_2},$$

or
$$\frac{\phi_1}{U} + \frac{\phi_2}{V} = 1. \quad \text{... (11)}$$

Dividing (*a*) by (*b*), we get

$$\frac{\phi_1}{U} \times \frac{V}{\phi_2} = -\frac{h_2}{h_1}.$$

$\therefore$ transverse magnification $\left(\frac{h_2}{h_1}\right) = -\frac{V\phi_1}{U\phi_2}$. ... (12)

Further, a ray PH_1 directed towards the first principal point H_1 must emerge from the second principal point H_2 along $H_2 P_1$ because H_1 H_2 and P, P_1 are parts of conjugate points.

Let θ_1 and θ_2 be the angles which the rays PH_1 and H_2P_1 make with the principal axis. Then applying Helmheltz Lagrange's law ($\mu_1 h_1 \tan\theta_1 = \mu_2 h_2 \tan\theta_2$) to the principal planes, we get

$$\mu_1 \tan\theta_1 = \mu_2 \tan\theta_2, \quad \text{... (c)}$$

since $h_1 = h_2$; the transverse magnification of the planes being h_2/h_1 is +1.

Substituting the values of $\tan\theta_1$ and $\tan\theta_2$ in equation (*c*) taking the conventional signs of U and V into consideration, we have

$$\mu_1 \frac{h_1}{-U} = \mu_2 \frac{-h_2}{V},$$

whence the lateral magnification

$$m = \frac{h_2}{h_1} = \frac{\mu_1}{\mu_2} \cdot \frac{V}{U} \quad \text{... (13)}$$

Comparing it with equation (12), we get

$$\frac{\phi_1}{\phi_2} = \frac{\mu_1}{\mu_2}. \quad \text{... (14)}$$

Further, combining relations (11) and (14), we get

$$-\frac{\mu_1}{\mu_2}\cdot\frac{\phi_2}{U}+\frac{\phi_2}{V} = 1,$$

or $$-\frac{\mu_1}{\mu_2}\cdot\frac{1}{U}+\frac{1}{V} = \frac{1}{\phi_2},$$

or $$\frac{\mu_1}{V}-\frac{\mu_1}{U} = \frac{\mu_2}{\phi_2}$$

$$= -\frac{\mu_1}{\phi_1}. \quad \text{... (15)}$$

For the lens system in air $\mu_1 = \mu_2$, so the relations (11), (13) and (14) can be written as

$$\frac{1}{V}-\frac{1}{U} = \frac{1}{\phi_2} = -\frac{1}{\phi_1}, \quad \text{... (16)}$$

$$m = \frac{h_2}{h_1} = \frac{V}{U}, \quad \text{... (17)}$$

and $$\phi_2 = -\phi_1. \quad \text{... (18)}$$

If equations (16), (17) and (18) are compared with the corresponding equations for a thin leas we find that coaxial lens system can be treated as a single unit like a thin lens provided the object and the image distances are measured from the corresponding principal points of the lens system.

Process of Transverse Magnification

For a combination of two lenses, the magnification is obtained by the ratio of the reduced image distance from the second principal plane to the reduced object distance from the first principal plane.

i.e., $$m = \frac{\text{reduced image distance}}{\text{reduced object distance}} = \frac{V}{U} \quad \text{... (8)}$$

Or, *alternatively*, if u_1 and v_1 are the distances of the object and the intermediate image from the first lens and u_2 and v_2

the distances of the intermediate image and the final image from the second lens, we write

$$\text{for first lens, } m_1 = \frac{v_1}{u_1}; \quad \text{for second lens, } m_2 = \frac{v_2}{u_2}.$$

From which magnification for the combination is given by

$$m = m_1 \times m_2 = \frac{v_1}{u_1} \times \frac{v_2}{u_2}. \quad \text{... (9)}$$

Significance of the Principal Points and Principal Foci : The principal points H_1, and H_2 of a lens system referred to in the preceding sections are a pair of conjugate points for which transverse magnification is plus unity or $m = +1$. This can be easily proved, where two rays, one originally parallel to the axis and the other through F_1, the first principal focus of the system are drawn in such a manner that the height above the axis of the incident parallel ray which passes on emergence through F_2, the second principal focus of the system, is the same as that of the emergent parallel ray which originated at F_1.

It is now obvious that the pair of rays converging to M_1, before entering the system becomes the pair of rays diverging from M_2 after emergence. According to our usual way of locating the image of an object point, the course of two rays from a given object point through the system to their final intersection on emergence, gives rise to the image point. Therefore, M_1 and M_2 are conjugate points of the system and the transverse planes M_1H_1 and M_2H_2 are conjugate planes. They are evidently planes of unit positive magnification, since the transverse magnification is unity, h_1 being equal to h_2. In other words these planes are such that a small object placed normal to the axis at H_1 will have an image equal to itself and similarly placed at H_2. They are called the unit planes or principal planes of the system. It is the points of intersection H_1 and H_2 of these planes with the principal axis that have been named as unit points (points of unit magnification) or the principal points of the system.

The positions of these points, of course, will depend upon the particular set of lenses. It is from these points that the distances may be measured in all cases, that is, the distance of the object from the first principal point and that of the image from the second principal point in order that the simple thin lens formula may be applied as usual. It is in fact for this reason that these points are called principal points and planes through them normal to the axis of the system principal planes. There is another important property of these points, *viz* , that when a lens system is situated in a uniform medium a ray directed towards the first point will, after refraction through the lens system, appear to proceed from the second point and parallel to the incident ray.

Just as for a single lens so for a lens system there are principal foci and principal focal planes. The first principal focus is at a distance F towards the left of the first principal point H_1, and the second principal focus is at the same distance towards the right of the second principal point H_2. They are generally denoted by F_1 and F_2 , respectively. The principal foci possess the usual properties, namely, that rays directed so as to pass through the first principal focus F_1 at incidence on the first lens will emerge through the second lens parallel to the axis and rays incident on the first lens parallel to the axis will emerge from the second lens in a direction passing through the second principal focus F_2. The planes passing through the principal foci and at right angles to the axis are called principal focal planes.

Numerically, the first focal length H_1F_1 and the second focal length H_2F_2 are equal, each being equal to F. But they are of opposite sign. According to our convention H_2F_2 is *positive* and H_1F_1 is *negative,*

i.e., $$H_2F_2 = -H_1F_1$$

The properties of the principal planes and the focal planes are frequently used in the treatment of eyepieces, in the complicated lens system such as are employed in the construction of objectives of high power microscopes.

The Principal Points of a Coaxial Lens System are Conjugated Foci: This can also be easily proved by considering the relation

$$\frac{1}{v-\beta} - \frac{1}{u-\alpha} = \frac{1}{F}.$$

By transposing, we have

$$\frac{1}{v-\beta} = \frac{1}{F} + \frac{1}{u-\alpha} = \frac{(u-\alpha)+F}{F(u-\alpha)},$$

or
$$v - \beta = \frac{F(u-\alpha)}{(u-\alpha)+F} \quad \text{... (10)}$$

When $u = \alpha$, $v - \beta = 0$ or $v = \beta$ for all values of F. Hence, an object at the first principal point will produce an image at the second principal point S.

Laws and Principles

Refraction of Light

When a ray of light passes from one transparent medium to another, in which it has a different velocity, there occurs a change in the direction of propagation of light except when it strikes the surface of separation of two media normally. This bending of a ray of light is called refraction. The angles made by the incident ray and the refracted ray with the normal to the separating surface at the point of incidence are called the angles, of incidence and of refraction , respectively. The refraction takes place according to the following laws:

Law 1. The incident ray, the normal and the refracted ray all lie in the same plane.

Law 2. The ratio of the sine of angle of incidence to the sine of angle of refraction for any two media is constant for a light of definite colour.

The second law was discovered by Willebrod Snellius in 1621 and is known after his name as Snell's Law.

Denoting the angles of incidence and refraction by i and r , respectively, the second law states that

$$\frac{\sin i}{\sin r} = {}^1\mu_2,$$

where μ is a constant depending upon the nature of media and the kind of light. It is called the refractive index of the second medium with respect to the first, the subscripts 1 and 2 indicating that the light passes from medium 1 to medium 2. If the direction of light is reversed, the ray will retrace its path, so that

$$\frac{\sin r}{\sin i} = {}^2\mu_1,$$

$$\therefore \quad {}^1\mu_2 \times {}^2\mu_1 = 1 \qquad \text{... (1)}$$

When a ray enters from vacuum into a medium, the above ratio is called the *absolute refractive index* of that medium.

The term refractive index also conveys a *physical idea*, as on the wave theory of light it expresses the ratio of the velocity of light in vacuum to its velocity in the medium, *viz.*,

$$\mu = \frac{\text{velocity of light in vacuum}}{\text{velocity of light in medium}} \qquad \text{... (2)}$$

Similarly, the refractive index of a medium B with respect to a medium A may be written as

$${}^A\mu_B = \frac{\text{velocity of light in medium } A}{\text{velocity of light in medium } B} \qquad \text{...(3)}$$

The refractive index of a medium varies with wavelength, optical density and temperature, etc.

Surface Power

The expression $\frac{(\mu_2 - \mu_1)}{r}$, is called the power of a refracting surface of refractive index μ_2 lying in a medium of refractive index μ_1. According to our convention the power of a convex surface is positive, since both $(\mu_1 - \mu_2)$ and r are positive while for a concave surface the power is negative because $(\mu_1 - \mu_2)$

is positive and r is negative. In general the power of a refracting surface which converges the incident parallel light is positive and that of a surface which diverges it, is negative. Denoting it by P, we have from equation (9).

$$P = \frac{\mu_2 - \mu_1}{r}$$

$$= \frac{\mu_2}{v} - \frac{\mu_1}{u}, \quad \text{... (11)}$$

or in case the first medium is air,

$$P = \frac{\mu - 1}{r} = \frac{\mu}{v} - \frac{1}{u}. \quad \text{... (11}a\text{)}$$

If r, u and v are expressed in metres, the power is expressed in diopters.

Principal Foci and Principal Focal Planes: Equation (9) gives us the position of two important points on the principal axis called the first and the second principal foci F_1 and F_2 of the surface , respectively. F_1 is the position of the object for which the image is at infinity. If we denote its distance from M by f_1, then by making v infinitely great, we have

$$\frac{\mu_2}{\infty} - \frac{\mu_1}{f_1} = \frac{\mu_2 - \mu_1}{r}$$

or

$$\frac{\mu_1}{f_1} = -\frac{\mu_2 - \mu_1}{r}$$

$$\therefore \quad f_1 = -\frac{\mu_1 r}{\mu_2 - \mu_1} \quad \text{... (12)}$$

If the first medium is air $\mu_1 = 1$, and $\mu_2 = \mu$ say, then

$$f_1 = \frac{r}{\mu - 1}. \quad \text{... (12}a\text{)}$$

Similarly, if the object is at infinity the position of the image is termed the second principal focus F_2. If f_2 be its distance from M, then

$$\frac{\mu_2}{f_2} - \frac{\mu_1}{\infty} = \frac{\mu_2 - \mu_1}{r},$$

or

$$\frac{\mu_2}{f_2} = \frac{\mu_2 - \mu_1}{r}$$

$$\therefore \quad f_2 = \frac{\mu_2 r}{\mu_2 - \mu_1} \qquad \ldots (13)$$

If the first medium be air, then

$$f_2 = \frac{\mu r}{\mu - 1} \qquad \ldots (13a)$$

An incident ray proceeding from the first principal focus of the surface gives rise to a refracted ray parallel to the axis and a ray parallel to the axis gives rise to a refracted ray proceeding towards the second principal focus of the surface.

Planes through F_1 and F_2, at right angles to the principal axis, are called the first and the second principal focal planes of the refracting surface.

From equations (12) and (13), we also get

$$\frac{f_1}{f_2} = -\frac{\mu_1}{\mu_2}$$

or

$$\frac{f_1}{\mu_1} + \frac{f_2}{\mu_2} = 0,$$

or

$$\mu_1 f_2 + \mu_2 f_1 = 0 \qquad \ldots (14)$$

If first medium is air

$$\mu_1 = 1$$

and

$$\mu_2 = \mu.$$

$$\therefore \quad f_2 + \mu f_1 = 0 \qquad \ldots (14a)$$

The distinction between the two focal distances should be carefully noted.

From (12*a*) $\quad -(\mu - 1) f_1 = r$

or numerically $\quad (\mu - 1) f_1 = r$

and from (14a) $\quad f_2 = -\mu f_1,$

or numerically $\quad f_2 - f_1 = \mu f_1 - f_1 = (\mu - 1) f_1$

That is, the numerical difference between the first and second focal distances is equal to the numerical value of the radius of curvature.

Also equation (9) can be expressed in the following forms:

(*i*) $$\frac{\mu_2}{v} - \frac{\mu_1}{u} = \frac{\mu_2 - \mu_1}{r} = -\frac{\mu_1}{f_1}, \left[\because f_1 = -\frac{\mu_1 r}{\mu_2 - \mu_1}\right]$$

$$\frac{\mu_2}{v} - \frac{\mu_1}{u} = \frac{\mu_2 - \mu_1}{r} = \frac{\mu_2}{f_2}. \quad \left[\because f_2 = -\frac{\mu_2 r}{\mu_2 - \mu_1}\right]$$

Either of these equations gives the conjugate distances for a single spherical surface and is analogous to the Gaussian formula for a thin lens.

(*iii*) Multiplying equation (9) by $\dfrac{r}{\mu_2 - \mu_1}$, we have

$$\frac{\mu_2 r}{(\mu_2 - \mu_1)v} - \frac{\mu_1 r}{(\mu_2 - \mu_1)u} = 1,$$

whence by equations (12) and (13)

$$\frac{f_2}{v} + \frac{f_1}{u} = 1, \quad \text{... (15)}$$

or $uv - vf_1 - uf_2 = 0$

and on adding $f_1 f_2$ to both sides of this equation, we get

$$(u - f_1)(v - f_2) = f_1 f_2, \quad \text{... (15}a\text{)}$$

a relation connecting u, v, f_1, and f_2.

Image of a Small Object by Refraction at a Spherical Surface: Consider a small object pP_1, in front of a convex spherical surface AMB. Let C be its centre of curvature, F_1 and F_2, the positions of the first and second principal foci. A ray parallel to the axis after refraction passes through the second principal focus F_2. The ray P_1C directed towards the centre of curvature goes undeviated and meets the other ray in Q_1 which is, therefore, the image of the point P_1. All other rays from P_1 will also be brought to focus at Q_1. For instance, the ray P_1F_1 which passes through the first principal focus will by definition of F_1 be refracted parallel to the axis and meet other rays in Q_1. The rest of the image lies in the conjugate plane through this point.

Transverse or Lateral Magnification: It is the ratio of any transverse dimension of the image to the corresponding transverse dimension of the object.

From right-angled triangles CPP_1 and CQQ_1

$$\frac{QQ_1}{PP_1} = \frac{CQ}{CP}$$

$$= \frac{MQ - MC}{MP + MC}$$

or
$$-\frac{h_2}{h_1} = \frac{v - r}{-u + r},$$

where
$$QQ_1 = MB = -h_2$$

and
$$PP_1 = AM = h_1$$

But $\frac{h_2}{h_1}$ is the *transverse magnification* denoted by m,

$$\therefore \qquad m = \frac{h_2}{h_1} = -\frac{v - r}{-u + r} = \frac{v - r}{u - r} \qquad \text{... (16)}$$

If m is positive, the image is virtual and erect, while if m is negative, the image is real and inverted.

Another useful expression for m is derived by making u se of the ray P_1M, which on refraction also meets at Q_1.

Such that $\angle P_1MP = \theta$ = angle of incidence

and $\angle Q_1MQ = \phi$ = angle of refraction.

Now $$\frac{QQ_1}{PP_1} = \frac{MQ \tan \phi}{MP \tan \theta}$$

or, *conventionally*

$$-\frac{h_2}{h_1} = \frac{v \tan \phi}{-u \tan \theta}.$$

$$\therefore \quad m = \frac{h_2}{h_1} = \frac{v}{u} \cdot \frac{\tan \phi}{\tan \theta}$$

$$= \frac{v}{u} \cdot \frac{\sin \phi}{\sin \theta}$$

$$= \frac{v}{u} \times \frac{\sin \phi}{\sin \theta} \qquad [\because \theta \text{ and } \phi \text{ are small}]$$

$$= \frac{v}{u} \frac{\mu_1}{\mu_2} \qquad \text{... [by Snell's law]}$$

$$= \frac{v}{\mu_2} \Big/ \frac{u}{\mu_1} \qquad \text{... (16a)}$$

where v/μ_2 and u/μ_1 are the reduced image and object distances in respective media.

If first medium is air, then $\frac{\mu_2}{\mu_1} = \mu$,

$$\therefore \quad m = \frac{v}{u\mu}. \qquad \text{... (16b)}$$

Magnification of Axis

It is the ratio of the depth of the image along the axis to the corresponding depth of the object.

Denoting it by m', we have

where u and v are the object and image distances from the surface.

Differentiating the image object relation

$$\frac{\mu_2}{v} - \frac{\mu_1}{u} = \frac{\mu_2 - \mu_1}{r}, \text{ we get}$$

$$\mu_2 \frac{dv}{v^2} - \mu_1 \frac{du}{u^2} = 0,$$

$$\text{and, therefore,} m' = \frac{dv}{du} = \frac{\mu_1}{\mu_2} \cdot \frac{v^2}{u^2}$$

$$= \frac{\mu_2}{\mu_1} \left(\frac{\mu_1 v}{\mu_2 u} \right)^2 = \frac{\mu_2}{\mu_1} m^2, \quad \text{... (17)}$$

where m is the transverse magnification. As $\mu_1 = 1$ for air then

$$m' = \mu m^2. \quad \text{... (17a)}$$

Also by differentiation of equation (15), we have

$$m' = \frac{dv}{du} = -\frac{v^2}{u^2} \cdot \frac{f_1}{f_2}. \quad \text{... (17b)}$$

Sine Condition of Abbe

It is a relation of great practical importance in Geometrical Optics and will be used later in deriving the magnifying power of a microscope.

Let QQ_1 be the image of an object PP_1 obtained by tracing the course of two known rays after refraction at a convex spherical surface separating a medium of refractive index μ_1 from a medium of refractive index μ_2. Let u, v, r denote the object distance, the image distance and the radius of curvature as usual.

An oblique ray PA making an angle α_1 with the axis meets the axis again at Q after refraction so that the refracted ray AQ makes an angle α_2 with the axis.

The line NAC it normal to the surface at A and let θ and ϕ be the angles of incidence and refraction , respectively. As regards the signs of the angles involved refer to conventions (4) and (5). According to them α_1, θ and ϕ, are positive and α_2 is negative.

From similar triangles CPP_1 and CQQ_1, we have

$$\frac{QQ_1}{PP_1} = \frac{OC}{PC}$$

$$= \frac{MQ - MC}{MP + MC},$$

or *conventionally* $-\dfrac{h_2}{h_1} = \dfrac{v - r}{-u + r},$

$$\text{or} \quad \frac{h_2}{h_1} = \frac{v - r}{u - r}, \qquad \dots (a)$$

where h_2 and h_1 are the respective heights of the image and the object.

Applying the law of sines to $\Delta PA\ C$, we have

$$\frac{\sin APC}{r} = \frac{\sin PAC}{-u + r},$$

$$\text{or} \quad \frac{\sin \alpha_1}{r} = \frac{\sin(180 - \theta)}{-u + r} = \frac{\sin \theta}{-(u - r)} \qquad \dots (b)$$

Similarly, from ΔQAC

$$-\frac{\sin \alpha_2}{r} = -\frac{\sin \alpha_2}{r}. \qquad \dots (c)$$

Dividing (*b*) by (*c*), we get

$$-\frac{\sin \alpha_1}{\sin \alpha_2} = \frac{\sin \theta}{\sin \phi} \times \frac{v - r}{-(u - r)},$$

$$\text{or} \quad \frac{\sin \alpha_1}{\sin \alpha_2} = \frac{\sin \theta}{\sin \phi} \times \frac{v - r}{u - r}$$

$$= \frac{\sin\theta}{\sin\phi} \times \frac{h_2}{h_1} \qquad \text{... [by equation } (a)]$$

$$= \frac{\mu_2}{\mu_1} \cdot \frac{h_2}{h_1},$$

$$\left[\because \quad \frac{\sin\theta}{\sin\phi} = \frac{\mu_2}{\mu_1} \text{ by Snell's law}\right]$$

whence $\mu_1 h_1 \sin\alpha_1 = \mu_2 h_2 \sin\alpha_2$. ... (18)

This relation is known as the sine condition. Since no where in deducing the result, the assumption of paraxial rays is introduced, the relation may be applied even if α_1 and α_2 are large, to any system of aplanatic surfaces, so that a point image is formed at Q for even wide-angled rays leaving P and passing through the system.

One important case of the application of this condition is that to the objective of a microscope.

The relation (18) can be written as

$$\frac{h_2}{h_1} \times \frac{\sin\alpha_2}{\sin\alpha_1} = \frac{\mu_1}{\mu_2}$$

Now, $\frac{h_2}{h_1}$ is called *linear transverse magnification* and $\frac{\sin\alpha_2}{\sin\alpha_1}$ represents *angular magnification.* So that, we have linear transverse magnification × angular magnification

$$= \frac{\mu_1}{\mu_2}.$$

For a single refracting surface of a small aperture, the sine condition becomes

$$\mu_1 h_1 \alpha_1 = \mu_2 h_2 \alpha_2 \qquad \text{... (18}a)$$

If there are $(n - 1)$ co-axial refracting surfaces, we have

$$\mu_1 h_1 \alpha_1 = \mu_2 h_2 \alpha_2 = \ldots.. = \mu_n h_n \alpha_n \qquad \text{... (18}b)$$

a relation often referred to as Helmholtz Lagrange's equation which may be stated as follows:

> "If a number of media are separated by spherical refracting surfaces with their centres on the same axis and are traversed by paraxial rays proceeding from a small object in one of the media then $\mu h\alpha$ is constant, where h denotes the size of the image formed in the medium of refractive index μ and α the angle which that part of a ray that lies in this medium makes with the axis".

Important Note: *In using the general relations deduced above and in the subsequent chapters on lenses for solving numerical problems it is necessary to substitute the numerical values of the data provided subject to the convention of signs used in deriving the formulae.*

Example: *A small object is placed 50 cm from a convex refracting surface of radius of curvature 20 cm. If the surface separates air from glass of refractive index 1.50, find (a) the power of the surface and (b) the position of the image.*

(*a*)
$$P = \frac{\mu_2 - \mu_1}{r} = \frac{1.50 - 1}{20/100}$$
$$= +\,0.50 \times 5 = +2.5\text{D}$$

(b)
$$\frac{\mu_2}{v} = \frac{\mu_1}{u} = \frac{\mu_2 - \mu_1}{r}.$$

Here,
$$u = -\,50 \text{ cm},$$
$$\mu_2 = 1.50,$$
$$\mu_1 = 1.$$

$$\therefore \quad \frac{1.50}{v} - \frac{1}{-50} = \frac{1.50 - 1}{20},$$

or
$$\frac{1.50}{v} + \frac{1}{50} = \frac{0.5}{20} = \frac{1}{40},$$

$$\text{or} \qquad \frac{1.50}{v} = \frac{1}{40} - \frac{1}{50}$$

$$= \frac{5-4}{200} = \frac{1}{200}.$$

$$\therefore \qquad v = 1.50 \times 200 = 300 \text{ cm}$$

$$= 3 \text{ m}$$

The positive sign indicates that the image lies on the positive side of the origin.

Example: *The incident face of a glass block is bounded by a concave surface of radius 3 cm. A small object 2 cm high is situated in air and on the axis 10 cm from the vertex. Find (a) the first and the second focal lengths, (b) the image distance, (c) the lateral magnificent and (d) the power of the surface. (R I. of glass = 150).*

$$(a) \qquad f_1 = \frac{\mu_1 r}{\mu_2 - \mu_1} = -\frac{1 \times (-3)}{1.50 - 1} = \frac{3}{0.50}$$

$$= 6.0 \text{ cm}$$

$$f_2 = \frac{\mu_2 r}{\mu_2 - \mu_1} = \frac{1.50 \times (-3)}{1.50 - 1} = \frac{-4.5}{0.50}$$

$$= -9.0 \text{ cm}$$

$$(b) \qquad \frac{\mu_2}{v} - \frac{\mu_1}{u} = \frac{\mu_2 - \mu_1}{r}.$$

$$\therefore \qquad \frac{1.50}{v} - \frac{1}{-10} = \frac{1.50 - 1}{-3}, \qquad [\because u = -10 \text{ cm}]$$

$$\text{or} \qquad \frac{1.50}{v} + \frac{1}{10} = \frac{0.50}{-3} = -\frac{1}{6}.$$

$$\text{or} \qquad \frac{1.50}{v} = -\frac{1}{6} - \frac{1}{10} = \frac{-5-3}{30} = -\frac{8}{30} = -\frac{4}{15}$$

$$(c) \qquad m = -\frac{1.50 \times 15}{4} = \mathbf{-5.625\,cm.}$$

$$m = \frac{\mu_1}{\mu_2} \times \frac{v}{u} = \frac{1}{1.50}\left(\frac{-5.625}{-10}\right) = \frac{5.625}{15}$$

$$= 0.375$$

$$\therefore \text{ size of image} = 0.375 \times \text{size of object}$$

$$= 0.375 \times 2$$

$$= 0.75 \text{ cm}$$

(*d*)
$$P = \frac{\mu_2 - \mu_1}{r} = \frac{1.50 - 1}{-3/100} = -\frac{0.50 \times 100}{3}$$

$$= -16.67 \text{ D}$$

Example: *A mark placed on the surface of a glass sphere is viewed through the glass from a portion directly opposite. If the diameter of the sphere is 10 cm and the refractive index of glass is 15, find the position of the image.*

The mark *P* on the surface of the glass sphere, is viewed through glass from the side *M*. The incident ray lies in the glass (μ_1 = 1.5) and is refracted at the surface *AM* into air (μ_2 = 1). Here, *conventionally* u = – 10 cm; r = – 5 cm. Applying the relation

$$\frac{\mu_2}{v} - \frac{\mu_1}{u} = \frac{\mu_2 - \mu_1}{r}, \text{ we have}$$

$$\frac{1}{v} - \frac{1.5}{-10} = \frac{1 - 1.50}{-5} = \frac{-0.50}{-5} = \frac{1}{10},$$

or
$$\frac{1}{v} = \frac{1}{10} - \frac{15}{100} = \frac{10 - 15}{100} = \frac{--5}{100},$$

$$\therefore \quad v = -20 \text{ cm}$$

that is, the image is formed at a distance of 20 cm from the surface *AM* and is virtual.

Example: *An object is placed 50 cm from the surface of a glass sphere of radius 12 cm along a diameter. Where will the final image* *be formed by refraction at both surfaces ? Refractive index of glass = 1.60.*

For refraction at first surface with vertex M, let v' be the distance of the image from it, then

$$\frac{\mu_2}{v'} - \frac{\mu_1}{u} = \frac{\mu_2 - \mu_1}{r_1}.$$

Here $\mu_1 = 1;\ \mu_2 = 1.60,$

$u = -\ 50$ cm.

$r_1 = +\ 12$ cm.

$$\therefore \quad \frac{1.60}{v'} - \frac{1}{-50} = \frac{1.60 - 1}{12} = \frac{1}{20}$$

or
$$\frac{1.60}{v'} = \frac{1}{20} - \frac{1}{50} = \frac{5-2}{100} = \frac{3}{100}$$

$$\therefore \quad v' = \frac{1.60 \times 100}{3} = \frac{160}{3} \text{ cm.}$$

This image serves as a virtual object for the second surface so that $(v' - 2r)$ becomes the object distance for the second surface.

For refraction at second surface,

$$\frac{\mu_2}{v} - \frac{\mu_1}{(v' - 2r)} = \frac{\mu_2 - \mu_1}{r_2}$$

Since $\mu_2 = 1;$

$\mu_1 = 1.60;$

$v' - 2r = (160/3) - 24 = +\ 88/3$ cm

and $r_2 = -\ 12$ cm.

$$\therefore \quad \frac{1}{v} - \frac{1.60}{\frac{88}{3}} = \frac{1 - 1.60}{-12} = \frac{1}{20}.$$

or $$\frac{1}{v} - \frac{4.80}{88} = \frac{1}{20}$$

or $$\frac{1}{v} = \frac{1}{20} + \frac{48}{880} = \frac{92}{880}$$

$$v = \frac{880}{92} = +9.5 \text{ cm}$$

The final image is real and 9.5 cm from M' on the side of the observer.

Law of Snell

Snell's law is sometimes expressed in a more symmetrical form deduced as follows:

Let μ_1 and μ_2 be the refractive indices for two media with respect to air, then

$$\mu_1 = \frac{\text{velocity of light in air}}{\text{velocity of light in medium (1)}} = \frac{C}{C_1}$$

and $$\mu_2 = \frac{\text{velocity of light in air}}{\text{velocity of light in medium (2)}} = \frac{C}{C_2}$$

where C is the velocity of light in air, C_1 and C_2 the respective velocities of light in the two media.

Dividing, we have

$$\frac{\mu_2}{\mu_1} = \frac{C_1}{C_2} = {}^1\mu_2$$

[by definition of refractive index]

By Snell's law

$${}^1\mu_2 = \frac{\sin i}{\sin r},$$

$$\therefore \quad \frac{\mu_2}{\mu_1} = \frac{\sin i}{\sin r},$$

or $$\mu_1 \sin i = \mu_2 \sin r.$$

That is, the product of refractive index and the sine of the angle made by the ray with normal at the point of incidence is constant for a given ray in both media.

Principle of Fermat

Optical Path and Fermat's Principle of Least Time: When a ray of light travels a distance S in a medium of refractive index μ, the product μS is said to be the optical path or the equivalent air path Since the value of μ for air does not appreciably differ from that for vacuum, it represents the distance in vacuum that light would travel in the same time in which it travels the distance S in the medium. .

When a ray travels distances S_1, S_2, S_3, etc. in media of indices μ_1, μ_2, μ_3, etc., then the optical path is given by

$$S = \mu_1 S_1 + \mu_2 S_2 + \mu_3 S_3 + \ldots\ldots$$
$$= \Sigma \mu_1 S_1.$$

For a medium of continuously varying optical density, the optical path of a ray PQ is expressed as an integral,

$$S = \int_P^O \mu dS. \quad \ldots(5)$$

Fermat's Principle: In its original form it was stated as follows:

When a ray of light goes from one point to another through a set of media, it always follows that path along which the time taken is minimum.

It should be noted that this original statement of the principle by Fermat is not complete and has been modified, for there are a number of cases where the optical path is *maximum* or else neither a maximum nor a minimum but *stationary* as in the case of formation of images by a lens in which all rays proceeding from an object point traverse the same *constant* path to the corresponding image point.

For a homogeneous and isotropic medium, μ being constant, we have

$$\int_P^Q dS = \text{a minimum, } i.e., \delta \int_P^Q dS = 0.$$

which means that between two points a ray will travel in a straight line and not any other path. Thus in such a medium *rectilinear* propagation of light is explained.

For a non-homogeneous and anisotropic medium

$$\int_P^Q \mu dS = \text{a minimum or a maximum or stationary,}$$

$$i.e., \text{ again } \delta \int_P^Q \mu dS = 0. \quad \dots (6)$$

This is the analytical expression of the second way of stating the law that *"for refraction through prisms and lenses the path of the ray between two points is that for which the time is* stationary *in value"*.

Derivation of Laws of Reflection and Refraction from Format's Principle

Law of Reflection: Let PA be the incident ray and AQ the reflected ray; i and r the angles of incidence and reflection , respectively. We have to show that the law of reflection, *viz.,*

$$\angle i = \angle r$$

is compatible with the principle of least time.

Let the perpendicular distances PL and QM be denoted by h_1 and h_2 , respectively and the total length on X-axis intercepted by these perpendiculars, *i.e., LM by p*.

The optical path PAQ is given by

$$S = PA + AQ$$

Taking $\qquad AM = x$

and $\qquad AL = (p - x)$, we get

$$S = \left[h_1^2 + (p-x)^2\right]^{\frac{1}{2}} + \left[h_2^2 + x^2\right]^{\frac{1}{2}}$$

Differentiating and equating the first derivative to zero, we have

$$\frac{dS}{dx} = \frac{1}{2}\frac{2(p-x)\times(-1)}{\left[h_1^2 + (p-x)^2\right]^{\frac{1}{2}}} + \frac{1}{2}\frac{2x}{\left(h_2^2 + x^2\right)^{\frac{1}{2}}} = 0,$$

$$\therefore \frac{p-x}{\left[h_1^2 + (p-x)^2\right]^{\frac{1}{2}}} = \frac{x}{\left(h^2 + x^2\right)^{\frac{1}{2}}}$$

i.e., $\qquad \sin i = \sin r$

or $\qquad i = r.$

Law of Refraction: PAQ is the path of a ray from the medium of refractive index μ_1 into a medium of refractive index μ_2 passing *via* a point A at the interface, then the optical path, PAQ is given by

$$S = \mu_1 PA + \mu_2 AQ.$$

Taking A as origin, $AM = x$ and $AL = (p - x)$, then

$$S = \mu_1\left[h_1^2 + (p-x)^2\right]^{\frac{1}{2}} + \mu_2\left[h_2^2 + x^2\right]^{\frac{1}{2}}$$

According to Fermat's principle S is either a maximum or a minimum or stationary.

In either case the first derivative of the expression above must vanish, that is

$$\frac{dS}{dx} = 0$$

Now $$\frac{dS}{dx} = \frac{\frac{1}{2}\mu_1}{\left[h_1^2+(p-x)^2\right]^{\frac{1}{2}}} \times (-2p+2x)$$

$$+\frac{\frac{1}{2}\mu_2}{\left(h_2^2+x^2\right)^{\frac{1}{2}}} \times 2x = 0,$$

$$\therefore \quad \mu_1 \frac{p-x}{\left[h_1^2+(p-x)^2\right]^{\frac{1}{2}}} = \frac{\mu_2 x}{\left(h_2^2+x^2\right)^{\frac{1}{2}}}$$

Since $$\frac{p-x}{\left[h_1^2+(p-x)^2\right]^{\frac{1}{2}}} = \sin i$$

and $$\frac{x}{\left[h_2^2+x^2\right]^{\frac{1}{2}}} = \sin r.$$

$$\therefore \quad \mu_1 \sin i = \mu_2 \sin r$$

The Snell's law is, therefore, proved.

Aplanatic Surface: An aplanatic surface has been defined, as one in which all rays originating from a point source P are brought to focus after reflection at the surface to some other point P'. These points P and P' are called aplanatic foci. These definitions hold for refracting surfaces as well. According to fermat's principle, the optical paths of all these rays must be equal so that if the points P and P' lie in different media the equation for the aplanatic surface becomes

$$\mu_1 PA + \mu_2 AP' = \text{constant} \qquad \ldots (7)$$

We can easily prove it for a single refraction. Let CD be an interface between two media of refractive indices μ_1 and μ_2

and PAP' the path of a ray from P to P' meeting the interface at A as we have to show that

$$\mu_1 PA + \mu_2 AP'$$

is stationary for the actual path. Let B be any point on CD close to A and let α_1 and α_2 be the complements of the angles PAB and $P'BA$, then obviously, α_1 the angle of incidence of the ray PA and α_2 the angle of refraction of the ray PB. But in the limit when AB is made infinitely small, α_2 is also the angle of refraction of the ray PA so that

$$\mu_1 \sin \alpha_1 = \mu_2 \sin \alpha_2$$

by Snell's law. Draw BL and AM perpendiculars on PA and BP'.

$$\text{Now } (\mu_1 PA + \mu_2 AP') - (\mu_1 PB + \mu_2 BP')$$

$$= \mu_1 (PA - PB) + \mu_2 (AP' - BP')$$

$$= \mu_1 AL - \mu_2 BM$$

$$= \mu_1 AB \cos \angle LAB - \mu_2 AB \cos \angle ABM$$

$$= \mu_1 AB \sin \alpha_1 - \mu_2 AB \sin \alpha_2$$

$$= AB[\mu_1 \sin \alpha_1 - \mu_2 \sin \alpha_2]$$

$$= 0, \text{ since } \mu_1 \sin \alpha_1 = \mu_2 \sin \alpha_2$$

as stated above.

$$\mu_1 PA + \mu_2 AP' = \mu_1 PB + \mu_2 BP'$$

Hence, the expression $\mu_1 PA + \mu_2 AP'$ is constant for small displacements of the point A and the locus of A is called an aplanatic surface. Such surfaces were first investigated by Descartes and are called *Cartesian ovals.*

In the case of reflection, the condition for the aplanatic surface becomes

$$PA + AP' = \text{constant, since } \mu_1 = \mu_2 \qquad \dots (8)$$

The properties of such a surface were briefly referred. For further details regarding aplanatic surfaces.

The application of Fermat's principle to refraction at a single surface and refraction through a lens is dealt with.

Refraction Through Spherical Surfaces: When a widely divergent pencil of rays is incident on a spherical refracting surface, separating two media, the refracted rays do not generally intersect in a single point but they touch a *caustic surface,* somewhat similar to that which occurs when reflection takes place at a spherical surface (Art. 1-6). If, however, a point be taken on the principal axis of the refracting surface and if we confine our attention to a very small centric pencil of rays incident on the surface, it will be found that after refraction the pencil will either converge to or appear to diverge from a single point.

Let us deduce relations connecting object and image distances in such cases. Here, we shall use the same general methods and conventions, etc. as employed in reflection from spherical surfaces.

Light Incident on a Convex Spherical Surface: Let *AMB* represent a convex spherical surface of radius of curvature r separating a less dense medium of refractive index μ_1 on the left from the more dense medium (shaded) of refractive index μ_2 on the right so that $\mu_2 > \mu_1$. Let C be the centre of curvature of the surface. Let P be a point source on the principal axis and PA a ray incident at A which after refraction, meets or appears to meet the axis in Q. θ and ϕ are the angles of incidence and refraction , respectively and α, β, γ are the angles which the incident ray, the refracted ray and the normal at A make with the principal axis. Then, by Snell's law,

$$\mu_1 \sin \theta = \mu_2 \sin \phi,$$

or if the angles are small

$$\mu_1\theta = \mu_2\phi. \quad \text{... } (a)$$

In ΔAPC, θ is the exterior angle, so that

$$\theta = \alpha + \gamma.$$

Similarly, for ΔAQC

$$\gamma = \phi + \beta,$$

or

$$\phi = \gamma - \beta.$$

Substituting the values of θ and ϕ in (*a*), we have

$$\mu_1 (\alpha + \gamma) = \mu_2 (\gamma - \beta),$$

Transposing, we get

$$\mu_2\beta + \mu_1\alpha = (\mu_2 - \mu_1)\,\gamma.$$

Denoting MA by y and applying the coordinate geometry conventions,

$$\alpha = \frac{y}{-u};$$

$$\beta = \frac{y}{v};$$

$$\gamma = \frac{y}{r}, \qquad [\because u \text{ is } -\text{ve}, v \text{ and } r \text{ are } +\text{ve}]$$

$\therefore$ from equation (*b*) $\mu_2 \dfrac{y}{v} + \mu_1 \dfrac{y}{-u} = (\mu_2 - \mu_1)\dfrac{y}{r}$,

$$\frac{\mu_2}{v} - \frac{\mu_1}{u} = \frac{\mu_2 - \mu_1}{r} \qquad \text{... (9)}$$

If the first medium be air, the formula reduces to

$$\frac{\mu}{v} - \frac{1}{u} = \frac{\mu - 1}{r}. \qquad \text{... (9a)}$$

We have

$$\theta = \alpha + \gamma$$

and

$$\phi = \beta + \gamma,$$

$$\therefore \quad \mu_1 (\alpha + \gamma) = \mu_2 (\beta + \gamma),$$

or

$$-\mu_2\beta + \mu_1\alpha = (\mu_2 - \mu_1)\gamma.$$

Substituting the values of α, β and γ with due regard being paid to conventional signs, we have

$$-\mu_2 \frac{y}{-v} + \mu_1 \frac{y}{-u} = (\mu_2 - \mu_1)\frac{y}{r}$$

$$\frac{\mu_2}{v} - \frac{\mu_1}{u} = \frac{\mu_2 - \mu_1}{r},$$

the same relation as (9) above.

Light Incident on a Concave Spherical Surface: Here the incident ray *PA* after refraction appears to intersect the axis in *Q*. With the same notation, we have as before,

$$\mu_1\theta = \mu_2\phi \quad \text{... } (a)$$

from above

$$\theta = \gamma - \alpha$$

and $$\phi = \gamma - \beta$$

Putting the values of θ and ϕ in (*a*) above, we have

$$\mu_1 (\gamma - \alpha) = \mu_2(\gamma - \beta).$$

Transposing

$$\mu_2\beta - \mu_1\alpha = (\mu_2 - \mu_1)\ \gamma. \quad \text{... } (b)$$

Now, $$\alpha = \frac{y}{-u},$$

$$\beta = \frac{y}{-v},$$

$$\gamma = \frac{y}{-r}$$

$\because$ u, v, r are all negative being on the left of origin,

$\therefore$ from equation (*b*)

$$\frac{\mu_2 y}{-v} - \frac{\mu_1 y}{-u} = (\mu_2 - \mu_1)\frac{y}{-r}$$

or on simplifying,

$$\frac{\mu_2}{v} - \frac{\mu_1}{u} = \frac{\mu_2 - \mu_1}{r},$$

or if the first medium be air, the formula becomes

$$\frac{\mu}{v} - \frac{1}{u} = \frac{\mu - 1}{r},$$

relations similar to (9) and (9*a*) above for a convex spherical surface.

For *large angles* of incidence and refraction and assuming the object and image distances u and v, respectively measured along the rays, the correct formula would become,

$$\frac{\mu_2}{v} - \frac{\mu_1}{u} = \frac{\mu_2 \cos \phi - \mu_1 \cos \theta}{r}.$$

It may be deduced for a concave surface as below:

According to our conventions

$$AP = -u;$$

$$AQ = -v$$

and $$AC = -r.$$

Now $$\Delta PAC = \Delta PAQ + \Delta QAC,$$

or trigonometrically,

$$\frac{1}{2} AP.AC \sin \theta = \frac{1}{2} AP.AQ \sin(\theta - \phi) + \frac{1}{2} AQ.AC \sin \phi.$$

or $$ur \sin \theta = uv \sin (\theta - \phi) + vr \sin \phi \qquad \text{... } (c)$$

Also, $$\frac{\sin \theta}{\sin \phi} = \frac{\mu_2}{\mu_1} \qquad \text{[Snell's law]}$$

Expanding Eqn. (*c*), we have

$$ur \sin \phi = uv \sin \theta \cos \phi$$
$$- uv \cos \theta \sin \phi + vr \sin \phi.$$

Dividing both sides by $uvr \sin\phi$ and rearranging, we have

$$\frac{1}{v}\frac{\sin\theta}{\sin\phi} = \frac{1}{r}\frac{\sin\theta}{\sin\phi}\cos\phi - \frac{1}{r}\cos\theta + \frac{1}{u}$$

Substituting $\frac{\mu_2}{\mu_1}$ for $\frac{\sin\theta}{\sin\phi}$, we get

$$\frac{1}{v}\frac{\mu_2}{\mu_1} = \frac{\cos\phi}{r}.\frac{\mu_2}{\mu_1} - \frac{1}{r}\cos\theta + \frac{1}{u},$$

or
$$\frac{\mu_2}{v} = \frac{\cos\phi}{r}.\mu_2 - \frac{\mu_1\cos\theta}{r} + \frac{\mu_1}{u},$$

or
$$\frac{\mu_2}{v} - \frac{\mu_1}{u} = \frac{\mu_2\cos\phi - \mu_1\cos\theta}{r} \qquad \text{... (10)}$$

if the first medium be air, we have

$$\frac{\mu}{v} - \frac{1}{u} = \frac{\mu\cos\phi - \cos\theta}{r} \qquad \text{... (10}a\text{)}$$

The derivation of a similar result for a convex surface is left as an exercise for the reader.

Conjugate Foci Relation for Refraction at a Spherical Surface by Fermat's Principle: Let C, be the centre of curvature of a convex spherical surface separating two isotropic media of refractive indices μ_1 and μ_2, respectively. Take M as the origin according to the coordinate geometry conventions adopted in the text. The equation to the section of the separating spherical surface in the plane of the diagram is

$$(x - r)^2 + y^2 = r^2$$

or
$$x^2 + y^2 = 2rx,$$

where $(r, 0)$ are the coordinates of C.

Let P $(u, 0)$ be the point object on the axis and $(v, 0)$ its image formed by refraction. Let PA be an incident ray at a point $A(x, y)$ on the surface and AQ the corresponding refracted ray. According to Fermat's law, the time taken from P to Q has an extreme value.

Now,

$$AP = \left[(x-u)^2 + y^2\right]^{\frac{1}{2}}$$

... [– ve sign being due to our convention

$$= -\left[x^2 + u^2 - 2xu + y^2\right]^{\frac{1}{2}}$$

$$= -\left[u^2 - 2xu + 2rx\right]^{\frac{1}{2}}$$

by the equation given above

$$= -u\left[1 - \left(\frac{2x}{u} - \frac{2rx}{u^2}\right)\right]^{\frac{1}{2}}$$

$$= -u\left[1 - \frac{x}{u} + \frac{rx}{u^2}\right]$$

$$= -\left[u - x + \frac{rx}{u}\right]$$

by binomial theorem, since all the terms except the first are small if paraxial rays alone are considered.

Similarly, $$AQ = \left[(v-x)^2 + y^2\right]^{\frac{1}{2}}$$

$$= v\left[1 - \frac{x}{v} + \frac{rx}{v^2}\right]$$

$$= \left[v - x + \frac{rx}{v}\right]$$

Now $\mu_1 AP + \mu_2 AQ$ = constant, by Fermat's principle. Then inserting the values of AP and AQ just obtained, we have

$$-\mu_1\left[u-x+\frac{rx}{u}\right]+\mu_2\left[v-x+\frac{rx}{v}\right] = \text{constant.}$$

Differentiating with respect to x, we get

$$-\mu_1\left[-1+\frac{r}{u}\right]+\mu_2\left[-1+\frac{r}{v}\right] = 0,$$

or
$$\mu_1-\frac{\mu_1 r}{u}-\mu_2+\frac{\mu_2 r}{v} = 0,$$

or
$$\frac{\mu_2 r}{v}-\frac{\mu_1 r}{u} = \mu_2-\mu_1,$$

whence
$$\frac{\mu_2}{v}-\frac{\mu_1}{u} = \frac{\mu_2-\mu_1}{r},$$

or if the first medium be air, we have

$$\frac{\mu}{v}-\frac{1}{u} = \frac{\mu-1}{r},$$

which are the usual refraction relations between conjugate foci.

By setting μ_1 for $-\mu_2$ and -1 for μ in the above obtain the relation we obtain the relation

$$\frac{1}{v}+\frac{1}{u} = \frac{2}{r},$$

giving the positions of conjugate foci in the case of reflection at a spherical surface.

Proceeding in a similar way we can derive these relations for a concave refracting surface. The reader is expected to attempt these solutions himself.

Permanent Laws

Laws that are Definite

When light travelling in a homogeneous medium is incident on a second medium, some of it is sent back into the first medium according to definite laws and is said to be reflected.

The angles which the incident ray and the reflected ray make with the normal to the surface are termed the angles of incidence and of reflection , respectively. The laws of reflection are as follows:

Law 1. The incident ray, the reflected ray and the normal to the surface, all lie in the same plane.

Law 2. The angle of incidence i is equal to the angle of reflection r.

The deviation suffered by the reflected ray is the angle by which its path has been deflected. It is equal to 180 – $(i + r)$. Since i and r are both equal, the deviation is given by $D = 180 - 2i$.

Spherical Nature

In deducing the general formulae for mirrors, we assumed that the angles between the incident rays or the reflected rays and the principal axis are small and a bundle of parallel rays close to the axis are brought to a sharp focus. This assumption is not justified when the aperture of the mirror is large. Rays that are reflected from a limited region of the mirror in the neighbourhood of the vertex M, are brought to focus at a point farther from the pole than marginal rays which cross the axis at points nearer the mirror.

This failure of rays to converge to or appear to diverge from a single point after reflection from a spherical mirror is called 'spherical aberration'. The effect of this is to destroy the sharpness of the image that would otherwise be formed by the mirror. Each ray after reflection is seen to converge and cut its neighbouring reflected ray at a point which lies on a curve called the caustic curve, the focus of the paraxial rays forming the cusp of the caustic.

We often see such a curve on the surface of tea in a cup, the light being reflected from the inside of the cup. It is the curve of greatest luminosity with its apex at F, the principal focus. The distance between the foci for the paraxial and marginal rays is taken as a measure of the axial spherical aberration. The rays reflected from the mirror are only drawn in one plane, *i.e.*, the plane of the paper.

Since the mirror is a part of a sphere, the caustic curve is an intersection by the plane of the paper of a surface which the reflected rays from all parts of the mirror touch. This surface is known as caustic surface. It is the surface traced out by the caustic curve, about the principal axis.

If a screen is placed at the paraxial focus F and then shifted towards the mirror, a point is reached where the size of the circular image is the *least*. This circular spot is called the circle of least confusion and may be taken as the nearest approach to the image of the distant object.

Aberration by Reflection at a Concave Surface: Let P be a point object lying on the axis MC of a concave mirror, with its centre of curvature at C and Q' be its image formed by reflection of a ray PA incident on the mirror at a point for removed from the axis. Let $\angle ACM = \gamma$, $PM = u$, $Q'M = v'$ $CM = r$ and $AN = h$ and it is such that h^3 and higher powers may be neglected in comparison with corresponding powers of u, etc.

Since AC bisects the angle PAQ'.

$$\frac{Q'C}{CP} = \frac{AQ'}{AP}$$

or $$Q'C.AP = CP.\ AQ' \quad \text{... (1)}$$

Now $$AP^2 = AC^2 + CP^2 - 2AC.\ CP.\ \cos(180 - \gamma)$$

[cosine formula]

$$= AC^2 + CP^2 + 2AC.CP\cos\gamma$$

$$= (-r)^2 + (-u+r)^2 + 2\times(-r)(-u+r)\cos\gamma\ .$$

$\because$ $$\cos\gamma = \sqrt{1-\sin^2\gamma}$$

$$= \left(1 - \frac{h^2}{r^2}\right)^{\frac{1}{2}} = 1 - \frac{h^2}{2r^2} \text{ approx.,}$$

$\therefore$ $$AP^2 = r^2 + (u-r)^2 + 2r(u-r)\left(1 - \frac{h^2}{2r^2}\right)$$

$$= u^2 - \frac{u-r}{r}h^2 \qquad \text{on simplification}$$

$$= u^2\left[1 - \frac{u-r}{ur}.\frac{h^2}{u}\right]$$

$$= u^2\left[1 - \left(\frac{1}{r} - \frac{1}{u}\right).\frac{h^2}{u}\right]$$

$$\therefore \qquad AP = u\left[1-\left(\frac{1}{r}-\frac{1}{u}\right).\frac{h^2}{u}\right]^{\frac{1}{2}}$$

$$= u\left[1-\left(\frac{1}{r}-\frac{1}{u}\right).\frac{h^2}{2u}\right]$$

by binomial expansion.

Similarly, $AQ' = v'\left[1-\left(\frac{1}{r}-\frac{1}{v'}\right).\frac{h^2}{2v'}\right]$

Also $\quad Q'C = (r - v')$

and $\quad CP = (u - r)$.

Substituting the values in equation (*i*), we get

$$(r-v')u\left[1-\left(\frac{1}{r}-\frac{1}{u}\right).\frac{h^2}{2u}\right] = (u-r)v'\left[1-\left(\frac{1}{r}-\frac{1}{v'}\right).\frac{h^2}{2v'}\right]$$

Dividing throughout by $uv'r$, we have

$$\left(\frac{1}{v'}-\frac{1}{r}\right)\left[1-\left(\frac{1}{r}-\frac{1}{u}\right).\frac{h^2}{2u}\right] = \left(\frac{1}{r}-\frac{1}{u}\right)\left[1-\left(\frac{1}{r}-\frac{1}{v'}\right).\frac{h^2}{2v'}\right]$$

Removing the brackets and simplifying, we get

$$\frac{1}{v'}+\frac{1}{u} = \frac{2}{r}+\left(\frac{1}{r}-\frac{1}{u}\right)\left(\frac{1}{v'}+\frac{1}{r}\right)\left(\frac{1}{v'}+\frac{1}{u}\right)\frac{h^2}{2} \quad .. (ii)$$

Since h^2 is small we can substitute v for v' in the terms forming the coefficient of h^2, where v is the image distance for paraxial rays.

$$\therefore \qquad \frac{1}{v'}+\frac{1}{u} = \frac{1}{v}+\frac{1}{u} = \frac{2}{r}$$

and the relation

$$\frac{1}{v'}+\frac{1}{u} = \frac{2}{r}$$ may be written as

$$\frac{1}{v'}-\frac{1}{r} = \frac{1}{r}-\frac{1}{u}$$

Therefore in the expression on the right-hand side of equation (*ii*) we can substitute.

$$\left(\frac{1}{r}-\frac{1}{u}\right)^2 \text{ for } \left(\frac{1}{r}-\frac{1}{u}\right)\left(\frac{1}{v'}-\frac{1}{r}\right) \text{ and } \frac{2}{r} \text{ for } \frac{1}{v'}+\frac{1}{u}.$$

Hence, equation (*ii*) becomes

$$\frac{1}{v'}+\frac{1}{u} = \frac{2}{r}+\left(\frac{1}{r}-\frac{1}{u}\right)^2 \frac{h^2}{r} \qquad \text{... } (iii)$$

$$\text{or} \quad \frac{1}{v'}-\left(\frac{2}{r}-\frac{1}{u}\right) = \left(\frac{1}{r}-\frac{1}{u}\right)^2 \frac{h^2}{r},$$

$$\text{or} \quad \frac{1}{v'}-\frac{1}{v} = \left(\frac{1}{r}-\frac{1}{u}\right)^2 \frac{h^2}{r} \qquad \left[\because \frac{2}{r}-\frac{1}{u}=\frac{1}{v}.\right]$$

$$\therefore \quad v'-v = \left(\frac{1}{r}-\frac{1}{u}\right)^2 \frac{h^2}{r} v^2 \qquad [\because v' \to v.]$$

If we put $h = a$, the semi-aperture of the mirror, we get

$$v'-v = \left(\frac{1}{r}-\frac{1}{u}\right)^2 \frac{a^2}{r} v^2$$

$$\text{or} \quad v'-v = \frac{\left(\frac{1}{r}-\frac{1}{u}\right)^2}{\frac{1}{v^2}} \times \frac{a^2}{r}$$

$$\text{But} \quad \frac{1}{v} = \frac{2}{r}-\frac{1}{u}$$

$$\therefore \quad v'-v = \frac{\left(\frac{1}{r}-\frac{1}{u}\right)^2}{\left(\frac{2}{r}-\frac{1}{u}\right)^2} \times \frac{a^2}{r} \qquad \text{... } (2)$$

Thus knowing u, r and a, $(v' - v)$ can be calculated.

The expression giving the value of $(v' - v)$ is a measure of the axial spherical aberration.

Astigmatism

If a narrow cone of light rays is incident *obliquely* on a small portion of a large concave mirror the reflected rays do not form a sharp focus even if the aperture is small. They are concentrated into a pair of broadened and mutually perpendicular lines called focal lines. This effect is known as astigmatism and the reflected pencil of rays is called astigmatic-pencil. Consider two rays PA and PB in the meridional or vertical plane diverging from a point P on the axis and incident on a small portion AB of the mirror making a large angle with the principal axis PM. Let these rays after reflection cut each other in T and meet the principal axis in Q and S, respectively. When A and B are very close together, T assumes a definite limiting position, *viz.*, the so-called *meridional focus*. Next consider the pencil with its principal ray PA proceeding along AQ after reflection. A neighbouring ray in the sagittal plane of the pencil would meet the mirror in A', a point near A and would after reflection pass through Q, since $PA'C$ is the plane containing the reflected ray. Then Q is the *sagittal focus*. The ΔPAB traces out a pencil of rays diverging from P. The point T traces out a short line TT' as shown perpendicular to the paper and the reflected rays pass through this line and also through the line SQ along the axis. The line TT' is called the first focal line F_1, and SQ the second focal line F_2 the light in this focal region forming a sort of *lemniscate*.

At a certain place in between the focal lines the length and the breadth of the patch of light are equal. This place is called the *circle of least confusion*, being the nearest approach to an image formed by the reflected pencil. The distance between the two focal lines is called astigmatic difference.

Expression for Astigmatic Difference: Where Q it the sagittal focus.

Let
$$AP = p,$$
$$AQ = q$$
and
$$AC = r,$$
$$\therefore \quad \Delta CAQ + \Delta PAC = \Delta\, PAQ$$

we have,

$$rq \sin\theta + rp \sin\theta = pq \sin 2\theta,$$

or
$$rq \sin\theta + rp \sin\theta = 2pq \sin\theta \cos\theta.$$

Dividing by $pqr \sin\theta$, we have

$$\frac{1}{p} + \frac{1}{q} = \frac{2\cos\theta}{r} \qquad \text{...(3)}$$

T is the meridional focus, if α, β and γ are the angles as indicated and $AT = q'$ then

$$\theta + \beta = \theta + \delta\theta + \gamma,$$

or
$$\delta\theta = \beta - \gamma$$

and
$$\theta + \gamma = \theta + \delta\theta + \alpha,$$

or
$$\delta\theta = \gamma - \alpha,$$

$$\therefore \quad \beta - \gamma = \gamma - \alpha$$

or
$$\alpha + \beta = 2\gamma. \qquad \text{... }(a)$$

Now,
$$\alpha = \frac{AB\cos\theta}{p},$$

$$\beta = \frac{AB\cos\theta}{q'},$$

$$\gamma = \frac{AB}{r}.$$

Putting these values in equation (*a*), we get

$$\frac{AB\cos\theta}{p}+\frac{AB\cos\theta}{q'}=\frac{2AB}{r},$$

whence $$\frac{1}{p}+\frac{1}{q'}=\frac{2}{r\cos\theta}. \quad \text{... (4)}$$

Subtracting equation (3) from equation (4), we get

$$\frac{1}{q'}-\frac{1}{q}=\frac{2}{r}\left[\frac{1}{\cos\theta}-\cos\theta\right]$$

$$=\frac{2}{r}\frac{1-\cos^2\theta}{\cos\theta}=\frac{2\sin^2\theta}{r\cos\theta}$$

$$=\frac{2}{r}\cos\theta\tan\theta,$$

or, the astigmatic difference

$$q-q'=\frac{2qq'}{r}\sin\theta\tan\theta \quad \text{...(5)}$$

Obviously, therefore, the astigmatic difference rapidly increases with increasing angle of incidence.

Note. Spherical aberration and Astigmatism can be reduced to a large extent by the use of a stop at the proper place so that the beam is incident at the central part of the mirror only.

The astigmatic effect for a parallel beam of rays originating in a point object at infinity and making a large angle with the principal axis *CM* of a concave mirror having a small aperture *AB*, centre C and polo *M*. Here again as before the rays give rise to two focal lines F_1 and F_2 separated by a finite distance and mutually perpendicular, so that even a mirror of small aperture will not produce a sharply focused image of a point object remote from its axis

From the above discussion we conclude that for oblique incidence or for wide-angled pencils a spherical mirror acts as if focal length were less than for direct pencils, the result being that a large object will give a curved image, which is also distorted, different parts being disproportionately magnified. Moreover, for points near the edge of the object large circles of confusion are produced and these cause considerable blurring of the image.

Mirrors: Ellipsoidal and Paraboloidal

To reduce spherical aberration we make use of ellipsoidal and paraboloidal mirrors. *CXD* is a section of the ellipsoidal mirror. The surface of such a mirror is one which is generated by an ellipse revolving about its major axis. The rays proceeding from a point source placed at one of the foci of the ellipse, say *P*, from, after reflection, a point image at the other focus *P′* no matter where they meet the mirror, the total length of the path of each ray between *P* and *P′* being always the same. The incident ray *PA* and the reflected ray *AP′* make equal angles with the normal drawn at *A* and so is the case with any other ray. Such a reflecting surface is called an aplanatic surface and is defined by the equation $PA + AP'$ =constant. The points *P* and *P′* are called aplanatic foci.

When one of the foci, either *P* or *P′* is at infinite distance as the aplanatic surface for reflection becomes a paraboloid of revolution. It is a more commonly used form of the mirror in which incident rays, parallel to the axis after reflection, are all concentrated at a single point as a consequence of the geometrical property of the parabola, *viz.*, the incident ray *AB*, parallel to the axis and the reflected ray *BS* are equally inclined to the normal at *B*, wherever the point *B* may be. There is, thus, a complete absence of spherical aberration, no matter how large the aperture of the mirror. This property has commended its use in the construction of high resolving power concave reflectors of the modern telescopes. The Mount Palomar telescope has a concave parapodial reflector with 5.08 m (200 inch) aperture.

Conversely, it may also be noted that a point source of light at the focus of a paraboloidal mirror will send out parallel rays after reflection from the surface of the mirror. It is for this reason that such mirrors are used in automobile headlights and searchlights.

Mirrors that are Spherical

Let us first of all assume that the surfaces are portions of a sphere. If the inside surface reflects, the mirror is concave and if it is the outside surface which reflects, it is convex.

The aperture of the mirror is the circular face presented to the incident tight and the width of the aperture is called its diameter. The centre of the aperture is called the pole of the mirror and the centre of the sphere of which the mirror forms a part is termed the centre of curvature. The line joining, the centre of curvature and the pole of the mirror is the principal axis. The laws of reflection stated above also hold in the case of spherical mirrors. However, before we proceed to determine the algebraic relations between the radius of curvature of a mirror, the distances of the object and the image from its pole we shall mention the convention of signs for distances and angles adopted in this book.

Convention of Signs: In deducing algebraic relations and expressions connecting distances of objects and images in all cases of reflection and refraction at spherical surfaces and in cases of lenses too, both converging and diverging, it is desirable to have a single formula applicable to both kinds of mirrors or surfaces or lenses, so it is necessary to adopt some sign conventions for *distances* and *angles* measured. Different writers have used different sign conventions, for each serves the purpose as well as any other. But to avoid confusion arising thereof the sign conventions which are now being largely used by most of the writers are those recommended by the International Commission on Optics.

1. Distances measured from the centre of mirror surface or lens in the direction in which light is travelling are

counted as 'positive' while those measured in the opposite direction are taken as 'negative'.

2. Transverse distances are counted as 'positive' when measured upwards from principal axis and 'negative' when measured below it.

3. The angles of incidence and reflection or refraction are taken as 'positive' when the radius of the surface must be rotated counter-clockwise through an angle less than $\pi/2$ to coincide with the direction of the ray.

4. The slope angles, *viz.*, the angles which the incident and reflected or refracted rays make with the axis are 'positive' when the axis of the surface must be rotated counter- clockwise through an angle less than $\pi/2$ to coincide with the direction of the ray.

It may be noted that these conventions agree with the usual Coordinate Geometry conventions for coordinates and positions so that if the centre of the mirror or lens or the curved surface is placed at the imaginary origin of the coordinates and the axis of the optical system coincides with the *X*-axis, the distances on the right of the origin are positive and on the left negative. The student is already familiar with the coordinate geometry conventions and therefore he shall have no difficulty in applying them readily to mirrors and lenses.

Formula of Mirror

Let *AMB* be the principal section of a spherical mirror with a small aperture and centre of curvature at

C. Let a ray *PA*, proceeding from a point source *P* lying off the axis, be reflected at *A* so that it makes equal angles with the normal *CA*, *i.e.*, $\angle i = \angle r$. The ray *PM* incident normally on the mirror reflects back on itself. These two rays intersect on the axis to form image *Q*. The mirror is concave and convex to incident light. Let the angles which *PA*, *QA* and *CA* make with the principal axis be denoted by α, β and γ , respectively. Then for the concave mirror

$$i = \gamma - \alpha$$

and $$r = \beta - \gamma$$

since $$i = r$$

$\therefore$ $$\beta - \gamma = \gamma - \alpha,$$

or $$\beta + \alpha = 2\gamma. \quad \text{... } (a)$$

Now $$\alpha = \frac{AM}{MP} = \frac{AM}{MQ}$$

and $$\gamma = \frac{AM}{MC}$$

If we denote AM, MP, MQ and MC by h, $-u$, $-v$ and $-r$, respectively, due regard being paid to the convention of signs,

then $$\alpha = \frac{h}{-u}, \; \beta = \frac{h}{-v}$$

and $$\gamma = \frac{h}{-r}$$

and equation *(a)* may be written as

$$\frac{h}{-v} + \frac{h}{-u} = \frac{2h}{-r}$$

or $$\frac{1}{v} + \frac{1}{u} = \frac{2}{r}.$$

For the convex mirror

$$i = \alpha + \gamma$$

and $$r = \beta - \gamma.$$

Since $$i = r, \; \therefore \alpha + \gamma = \beta - \gamma,$$

or $$\beta - \alpha = 2\gamma. \quad \text{... } (b)$$

Now $$\alpha = \frac{AM}{MP} = \frac{h}{-u};$$

$$\beta = \frac{AM}{MQ} = \frac{h}{v}$$

and $$\gamma = \frac{AM}{MC} = \frac{h}{r}.$$

Putting these values in equation (*b*) above, we have

$$\frac{h}{v} - \frac{h}{-u} = \frac{2h}{r},$$

or $$\frac{1}{v} + \frac{1}{u} = \frac{2}{r}. \quad \text{... (1)}$$

This is the same result as obtained above for a concave mirror.

It may be noted that this equation is independent of the angles of incidence, provided the aperture is *small*, so that the position of Q is the same for the given position of P for all values of the angle of incidence.

If the object is at infinite distance from the mirror, the particular point at which image is formed is called the focus of the mirror and the distance of the point from M is called the focal length of the mirror, *i.e.*,

$$f = \underset{u \to \infty}{\text{Lt.}} [v]$$

Substituting in (1), we get

$$\frac{1}{f} + \frac{1}{\infty} = \frac{2}{r} \text{ or } f = \frac{r}{2}.,$$

that is, focal length is equal to half the radius of curvature, the formula (1) thus becomes

$$\frac{1}{v} + \frac{1}{u} = \frac{1}{f} \quad \text{... (1}a\text{)}$$

The points P and Q connected by this relation are known as conjugate points and transverse planes through these points are called conjugate planes.

Important Note: *In using the relation*

$$\frac{1}{v}+\frac{1}{u} = \frac{1}{f}$$

in the solution of numerical problems, the student must never forget to substitute in it the conventional signs for the quantities whose values are given. No sign is to be attached to the quantity to be determined.

Focal Plane: A plane perpendicular to the axis and drawn through the focus is called focal plane. Incident parallel rays making some angle with the principal axis are brought to focus at some point in the focal plane.

The image Q of a distant point object, sending an oblique pencil of parallel rays is formed at the intersection with focal plane of that ray which goes through the centre of curvature.

Bibliography

Abramowitz, Mortimer: *Contrast Methods in Microscopy: Transmitted Light,* Olympus America, New York, 1987.

————————: *Optics: A Primer,* Olympus America, New York, 1994.

————————: *Photomicrography: A Practical Guide,* Olympus America, New York, 1998.

Allan, Victoria J.: *Protein Localization by Fluorescence Microscopy: A Practical Approach,* Oxford University Press, Oxford, 2000.

Anderson, M. D.: *Through the Microscope,* The Natural History Press, New York, 1965.

Archinard, P. E.: *Microscopy and Bacteriology: A Manual for Students and Practitioners,* Lea Brothers and Company, Philadelphia, 1903.

Ash, Eric A.: *Scanned Image Microscopy,* Academic Press, London, 1980.

Avinoam, Tomer: *Structure of Metals through Optical Microscopy,* ASM International, Ohio, 1991.

Bach, Peter H. and Poole Phil L.: *Biotechnology Applications of Microinjection, Microscopic Imaging and Fluorescence,* Plenum Press, New York, 1993.

Bainbridge, D.: *Extensible Optical Music Recognition,* University of Canterbury, Christchurch, 1997.

——————: *Preliminary Experiments in Musical Score Recognition*, University of Edinburgh, Edinburgh, 1991.

Baker, Henry and Wilson James: *The Microscope Made Easy and Pocket Microscopes*, Science Heritage Ltd., Lincolnwood, 1987.

Baker, John R.: *Cytological Technique: The Principles Underlying Routine Methods*, Methuen and Company, London, 1966.

Barer, R.: *Lecture Notes on the Use of the Microscope*, Blackwell Scientific Publications, Oxford, 1956.

Barnard, Joseph Edwin: *Practical Photo-Micrography*, Edward Arnold and Company, London, 1911.

Bass, Michael David R. and Wolfe William L.: *Handbook of Optics: Devices, Measurements and Properties*, Optical Society of America, McGraw-Hill, New York, 1995.

Bausch, Edward: *Manipulation of the Microscope*, Bausch and Lomb Optical Company, New York, 1891.

——————: *The Use and Care of the Microscope*, Bausch and Lomb Optical Company, New York, 1932.

Beale, Lionel S.: *How to Work with the Microscope?*, Harrison, Pall Mall, London, 1868.

Beran, T.: *Rozpoznavani Notoveho Zapisu (In Czech)*, Czech Technical University, Czech Republic, 1997.

Blostein, D. and Baird H. S.: *A Critical Survey of Music Image Analysis*, Springer-Verlag, Berlin, 1992.

Caduto, Michael J.: *Pond and Brook: A Guide to Nature in Fresh Water Environments*, University Press of New England, New Hampshire, 1990.

Carter, N. P.: *A Generalized Approach to Automatic Recognition of Music Scores: Department of Music*, Stanford University, 1993.

Carucci, John: *The New Media Guide to Creative Photography: Image Capture and Printing in the Digital Age,* Amphoto Books, New York, 1988.

Castleman, Kenneth R.: *Digital Image Processing,* Prentice-Hall, New Jersey, 1996.

Catling, Dorothy and Grayson John: *Identification of Vegetable Fibres,* Archetype Publications, London, 1998.

Cebulla, D. W.: *Handbook of Incident Light Microscopy,* Carl Zeiss, Germany, 1979.

Cellis, Julio E.: *Cell Biology: A Laboratory Handbook,* Academic Press, California, 1998.

Chalfie, Martin and Kain Steven: *Green Fluorescent Protein: Properties, Applications and Protocols,* John Wiley and Sons, New York, 1998.

Chamot, Emile Monnin: *Elementary Chemical Microscopy,* John Wiley and Sons, New York, 1921.

Choi, J.: *Optical Recognition of the Printed Musical Score, (Electrical Engineering and Computer Science),* University of Illinois at Chicago, 1991.

Davies, Adrian and Fennessy Phil: *Digital Imaging for Photographers,* Focal Press, Oxford, 1998.

Davies, Thomas: *The Preparation and Mounting of Microscopic Objects,* David Bogue, London, 1880.

DeDuve, Christian: *Blueprint for a Cell: The Nature and Origin of Life,* Neil Patterson Publishing, North Carolina, 1991.

DeHoff, Robert T. and Rhines Frederick N.: *Quantitative Microscopy,* McGraw-Hill Book Company, New York, 1968.

Delly, John Gustav: *Photography through the Microscope,* Eastman Kodak Company, New York, 1988.

————————: *Teaching Microscopy,* Microscope Publications Ltd., Chicago, 1994.

Desmarias, Louis: *Applied Electro Optics*, Prentice Hall PTR, New Jersey, 1998.

Diaspro, Alberto: *Confocal and Two-Photon Microscopy: Foundations, Applications and Advances*, John Wiley and Sons, New York, 2002.

Diener, G. R.: *Modeling Music Notation: A Three-dimensional Approach*, Stanford University, 1990.

Dimitrov, L.: *Computergraphische Darstellung Kinematischer Probleme*, Technische Universität, Wien, 1984.

Disraeli, Robert: *New Worlds Through the Microscope*, The Viking Press, New York, 1960.

Dobell, Clifford: *Antony Van Leeuwenhoek and His 'Little Animals': Being an Account of the Father of Protozoology and Bacteriology and His Multifarious Discoveries in These Disciplines*, Russell and Russell, New York, 1958.

Doggart, James Hamilton: *Ocular Signs in Slit-Lamp Microscopy*, Henry Kimpton, London, 1948.

Dougherty, E. R.: *Electronic Imaging Technology*, SPIE Press, Washington, 1999.

Drew, Aubrey H. and Wright Lewis: *The Microscope: A Practical Handbook*, The Religious Tract Society, London, 1922.

Echlin, Patrick: *Low-Temperature Microscopy and Analysis*, Plenum Press, New York, 1992.

Edmund Scientific: *Collimators and Collimation*, Edmund Scientific Company, New Jersey, 1998.

Edmund, N.W. and Brown Sam: *The Optical Bench*, Edmund Scientific Company, New Jersey, 1996.

Edwards, Frank B. and Aziz Laurel: *Close Up: Microscopic Photographs of Everyday Stuff*, Bungalo Books, Canada, 1992.

Efford, Nick: *Digital Image Processing: A Practical Introduction Using Java*, Pearson Education Ltd., United Kingdom, 2000.

Ekstrom, Michael P.: *Digital Image Processing Techniques*, Academic Press, Florida, 1983.

Elmore, William C. and Heald Mark A.: *Physics of Waves*, Dover Publications, New York, 1969.

Ernst Leitz: *Image-Forming and Illuminating Systems of the Microscope*, Ernst Leitz, Ltd., Canada, 1970.

Fahmy, H.: *A Graph-grammar Approach to High-level Music Recognition*, Queen's University, Canada, 1991.

Falk, David and Stork David: *Seeing the Light: Optics in Nature, Photography, Color, Vision and Holography*, John Wiley and Sons, New York, 1986.

Fillard, J. P.: *Near Field Optics and Nanoscopy*, World Scientific Publishing Company, Singapore, 1997.

Fischer, K. N.: *Computer Recognition of Engraved Music*, University of Tennessee, 1978.

Fischer, Robert E. and Tadic-Galeb Biljana: *Optical System Design*, SPIE Press, New York, 2000.

Fleming, Denise: *In the Small, Small Pond*, Harry Holt and Company, New York, 1993.

Ford, Brian J.: *Single Lens: The Story of the Simple Microscope*, Harper and Row Publisher, Cambridge, 1985.

————————: *The Revealing Lens*, George G. Harrap and Company Ltd., London, 1973.

Gage, Simon Henry: *The Microscope: An Introduction to Microscopic Methods and to Histology*: Comstock Publishing Company, New York, 1920.

Gander, Ralph: *Photomicrographic Technique for Medical and Biological Scientists*, Hafner, New York, 1969.

Gardener, Robert: *Experiments with Light and Mirrors*, Enslow Publishers, New Jersey, 1995.

Garnett, W. J.: *Freshwater Microscopy*, Constable and Company Ltd., London, 1953.

Geil, Phillip H.: *Polymer Single Crystals*, Interscience Publishers, New York, 1963.

Georges, Gregory: *Edit, Use and Share Your Digital Images: Digital Camera Solutions*, Muska and Lipman Publishing, Ohio, 2000.

Gifkins, R. C.: *Optical Microscopy of Metals*, American Elsevier, New York, 1970.

Goldstein, D. J.: *Understanding the Light Microscope: A Computer-Aided Introduction*, Academic Press, London, 1999.

Gonzalez, Rafael C. and Woods Richard E.: *Digital Image Processing*, Addison-Wesley Publishing Company, Massachusetts, 1993.

Hall, Charles A. and Linssen E. F.: *How to Use the Microscope?*, A and C Black Ltd., London, 1968.

Hallimond, A. F.: *The Polarizing Microscope*, Vickers Ltd., United Kingdom, 1970.

Halpern, Alvin M. and Erlbach Erich: *Schaum's Outline of Theory and Problems of Beginning Physics: Waves, Electromagnetism, Optics and Modern Physics*, McGraw-Hill, New York, 1998.

Hamashima, Yoshiro: *The Use of the Olympus Fluorescence Microscope.*, Olympus Optical Company, Tokyo, 1982.

Harrison, William C.: *The Microscope*, Julian Messner, New York, 1965.

Hartley, W. G.: *The Light Microscope: Its Use and Development*, Senecio Publishing Company, United Kingdom, 1993.

——————: *The Microscope: A Basic Guide*, The Quekett Microscopical Club, London, 1983.

Hartshorne, Norman H.: *The Microscopy of Liquid Crystals*, Microscope Publications Ltd., United Kingdom, 1974.

Inoue, Shinya: *Video Microscopy*, Plenum Press, New York, 1986.

Isenberg, G.: *Modern Optics, Electronics and High Precision Techniques in Cell Biology*, Springer-Verlag, Germany, 261 pages (1998).

Ivens, Virginia R. and Levine Norman D.: *Principal Parasites of Domestic Animals in the United States: Biological and Diagnostic Information*, University of Illinois at Urbana-Champaign, Urbana, 1978.

Jacker, Corinne: *Window on the Unknown: A History of the Microscope*, Charles Scribner's Sons, New York, 1966.

Jackson, Alan: *Amateur Photomicrography with Simple Apparatus*, The Focal Press, London, 1948.

Jaffe, Howard W: *Crystal Chemistry and Refractivity*, Dover Publications, New York, 1996.

Jahn, Theodore Louis: *How to Know the Protozoa?*, McGraw-Hill, Boston, Massachusetts, 1949.

Jain, Anil K.: *Fundamentals of Digital Image Processing*, Prentice-Hall, New Jersey, 1989.

James, J. and Tanke H. J.: *Biomedical Light Microscopy*, Kluwer Academic Publishers, Dordrecht, Netherlands, 1991.

Janesick, James R.: *Scientific Charge-Coupled Devices*, SPIE Press, Washington, 2001.

Javois, Lorette C.: *Immunocytochemical Methods and Protocols*, Humana Press, New Jersey, 1999.

Jenkins, Francis A. and White Harvey E.: *Fundamentals of Optics*, McGraw-Hill, New York, 1976.

Kasten, F. H.: *Cell Structure and Function by Microspectrofluorometry*, Academic Press, New York, 1989.

Kavanaugh, James: *Pond Life: An Introduction to Familiar Plants and Animals Living on or Near Ponds, Lakes and Wetlands*, Waterford Press Ltd., Washington, 1997.

Kawata, Satoshi: *Near-Field Optics and Surface Plasmon Polaritons*, Springer-Verlag, Berlin, 2001.

Keeney, Arthur H. and Fratello Cosmo J.: *Dictionary of Ophthalmic Optics*, Butterworth-Heinemann, Massachusetts, 1995.

Kehl, George L.: *The Principles of Metallographic Laboratory Practice*, McGraw-Hill, New York, 1949.

Kerr, Richard: *Nature through Microscope and Camera*, The Religious Tract Society, United Kingdom, 1909.

King, Julie Adair: *Digital Photography for Dummies*, IDG Books Worldwide, California, 1998.

Lackie, John M. and Dow Julian A.T.: *The Dictionary of Cell and Molecular Biology*, Academic Press, London, 1999.

Lakowicz, Joseph R.: *Nonlinear and Two-Photon-Induced Fluorescence*, Plenum Press, New York, 1997.

Lakowicz, Joseph R.: *Principles of Fluorescence Spectroscopy*, Kluwer Academic/Plenum Publishers, New York, 1999.

——————: *Protein Fluorescence*, Plenum Publishers, New York, 2000.

Lamb, Trevor and Bourriau Janine: *Colour: Art and Science*, Cambridge University Press, Cambridge, 1995.

Landolt, H.: *Handbook of the Polariscope and its Practical Applications*, MacMillan and Company, London, 1882.

Madsen, Christi K. and Zhao Jian H.: *Optical Filter Design and Analysis: A Signal Processing Approach*, John Wiley and Sons, New York, 1999.

Malies, Harold M.: *A Short History of the English Microscope*, Microscope Publications Ltd., Chicago, 1981.

——————: *Applied Microscopy and Photomicrography*, Fountain Press, United Kingdom, 1959.

Mann, Allen: *Infrared Optics and Zoom Lenses*, SPIE Press, Washington, 2000.

Marchand-Maillet, S. and Sharaiha Y. M.: *Binary Digital Image Processing: A Discrete Approach*, Academic Press, California, 2000.

Nachtigall, Werner: *Exploring with the Microscope: A Book of Discovery and Learning*, Sterling Publishing Company, New York, 1995.

Nassau, Kurt: *The Physics and Chemistry of Color: The Fifteen Causes of Color*, John Wiley and Sons, New York, 2001.

Needham, George Herbert: *Practical Use of the Microscope, Including Photomicrography*, Charles C. Thomas Publishers, Illinois, 1958.

Needham, George Herbert: *The Microscope: A Practical Guide*, Charles C. Thomas Publishers, Illinois, 1968.

Needham, James G. and Needham Paul R.: *A Guide to the Study of Fresh-Water Biology*, McGraw-Hill, Massachusetts, 1962.

Nesse, William D.: *Introduction to Optical Mineralogy*, Oxford University Press, New York, 1991.

Ogle, Robert R.: *Atlas of Human Hair: Microscopic Characteristics*, CRC Press, Florida, 1999.

Ohtsu, Motoichi and Hori Hirokazu: *Near-Field Nano-Optics: From Basic Principles to Nano-Fabrication and Nano-Photonics*, Plenum Publishers, New York, 1999.

Ohtsu, Motoichi: *Near-Field Nano/Atom Optics and Technology*, Springer-Verlag, Japan, 1998.

Palmer, F. W. and Sahiar A.B.: *Microscopes to the End of the Nineteenth Century*, Her Majesty's Stationary Office, United Kingdom, 1971.

Pantin, C. F. A.: *Notes on Microscopical Technique for Zoologists*, Cambridge University Press, Cambridge, 1946.

Parker, J. R.: *Algorithms for Image Processing and Computer Vision*, John Wiley and Sons, New York, 1997.

Payne, B.O.: *Microscope Design and Construction*, Troughton and Simms Ltd., United Kingdom, 1954.

Peacock, Alan H.: *Elementary Microtechnique*, Edward Arnold and Company, London, 1940.

Pendergrass, William R.: *Carolina Protozoa and Invertebrates Manual*, Carolina Biological Supply Company, North Carolina, 1980.

Quekett, John: *Practical Treatise on the Use of the Microscope*, Science Heritage Ltd., Lincolnwood, 1987.

Radley, J. A. and Grant Julius: *Fluorescence Analysis in Ultra-Violet Light*, Chapman and Hall Ltd., London, 1959.

Ratledge, David: *The Art and Science of CCD Astronomy*, Springer-Verlag, London, 1997.

Ray, Sidney F.: *Applied Photographic Optics*, Focal Press, Oxford, 1997.

Rhodes, Marion B.: *Selected Papers on Optical Microscopy*, SPIE Optical Engineering Press, Washington, 2000.

Richards, Oscar W.: *The Effective Use and Proper Care of the Microscope*, American Optical Company Instrument Division, New York, 1958.

Richardson, James H.: *Handbook for the Light Microscope: A User's Guide*, Noyes Publications, New Jersey, 1991.

Saggerson, E.P.: *A Handbook of Minerals under The Microscope*, University of Natal Press, Pietermaritzburg, 1986.

Sandon, H.: *Essays on Protozoology*, Hutchinson Educational, United Kingdom, 1968.

Savage, Candace: *Aurora: The Mysterious Northern Lights*, Sierra Club Books, California, 1995.

Schade, Karl-Heinz: *Light Microscopy: Technology and Application*, Verlag Moderne Industrie, Germany, 1994.

Schwarz, Edward Robinson: *Textiles and the Microscope*, McGraw-Hill Book Company, New York, 1934.

Scott, Craig: *Introduction to Optics and Optical Imaging*, IEEE Press, New Jersey, 1998.

Shannon, Robert R.: *The Art and Science of Optic Design*, Cambridge University Press, Cambridge, 1997.

Stransky, G. Dimitrov, L. Wenger, E. and Eberl R.: *Collagen Fibrils in Idiopathic Carpal Tunnel Syndrome*, European Journal of Cell Biology, 1987.

Taylor, D. Lansing and Wang Yu-Li: *Fluorescence Microscopy of Living Cells in Culture*, Academic Press, New York, 1989.

Thomson, D. and Bradbury, S.: *An Introduction To Photomicrography*, Oxford Scientific Publications, Oxford, 1987.

Tobias, J. Carroll: *The Student's Manual of Microscopic Technique*, American Photographic Publishing Company, Massachusetts, 1936.

Traber, H. A.: *The Microscope as a Camera*, Focal Press, London, 1971.

Uytenbogaardt, W. and Burke E. A. J.: *Tables for Microscopic Identification of Ore Minerals*, Dover Publications, New York, 1985.

VanCleave, Janice: *Microscopes and Magnifying Lenses*, John Wiley and Sons, New York, 1993.

Viney, Christopher: *Transmitted Polarized Light Microscopy*, McCrone Research Institute, Illinois, 1990.

Wade, Nicolas J.: *A Natural History of Vision*, The MIT Press, Cambridge, Massachusetts, 1998.

Walker, Bruce H.: *Optical Engineering Fundamentals,* SPIE Optical Engineering Press, Washington, 1998.

Walker, M.I.: *Amateur Photomicrography,* Chilton Book Company, Philadelphia, Pennsylvania, 1972.

Walter, Friedrich: *The Microtome: Manual of the Technique of Preparation and of Section Cutting,* Ernst Leitz Ltd., Canada, 1983.

Weis, S. Stransky G. Dimitrov L. and Eberl R.: *Proceedings of the Austrian-Hungarian Joint Conference on Electron Microscopy,* Seggau, Austria, 1987.

Wolfe, William L.: *Handbook of Optics: Fundamentals, Techniques and Design,* Irregular Pagination, New York, 1995.

Yates, Raymond F.: *Exploring with the Microscope,* Appleton-Century Company, New York, 1934.

Yuste, Rafael and Konnerth Arthur: *Imaging Neurons: A Laboratory Manual,* Cold Spring Harbor Press, New York, 1999.

Zhu, Xing and Ohtsu Motoichi: *Near-Field Optics: Principles and Applications,* World Scientific Publishing, Singapore, 2000.

Zieler, H. Wolfgang: *The Optical Performance of the Light Microscope,* Microscope Publications, Chicago, 1972.

Zirkel, Ferdinand: *Microscopical Petrography,* United States Government Printing Office, Washington, 1876.

Index

N

O

P

R

S

T

U

W

❑❑❑